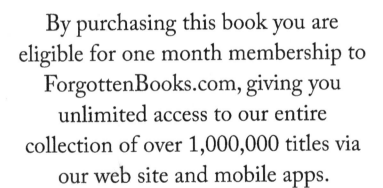

ISBN 978-0-365-03973-0
PIBN 10045247

A NATURAL HISTORY

OF THE GLOBE, OF MAN, OF BEASTS, BIRDS, FISHES, REPTILES, INSECTS AND PLANTS.

FROM

THE WRITINGS OF BUFFON, CUVIER AND OTHER EMINENT NATURALISTS.

A NEW EDITION,

WITH MODERN IMPROVEMENTS, AND FIVE HUNDRED ENGRAVINGS.

BOSTON:

GRAY AND BOWEN.

MDCCCXXXI.

NATURAL HISTORY

OF

THE GLOBE, OF MAN, OF BEASTS, BIRDS, FISHES, REPTILES, INSECTS AND PLANTS.

FROM

THE WRITINGS OF BUFFON, CUVIER, LACEPEDE,

AND OTHER EMINENT NATURALISTS.

EDITED BY JOHN WRIGHT,

MEMBER OF THE ZOOLOGICAL SOCIETY OF LONDON.

A NEW EDITION,

WITH IMPROVEMENTS FROM GEOFFREY, GRIFFITH, RICHARD-
SON, LEWIS AND CLARK, LONG, WILSON, AND OTHERS.

WITH FIVE HUNDRED ENGRAVINGS.

IN FIVE VOLUMES.

VOL. IV.

BOSTON:
PUBLISHED BY GRAY AND BOWEN.

1831.

ENTERED according to Act of Congress, in the year 1831,
By SAMUEL G. GOODRICH,
In the Clerk's Office of the District Court of Massachusetts.

CONTENTS OF VOL. IV.

CHAP. VI.

CHAP. VII.

CHAP. VIII.—PART I.

CHAP. VIII.—PART II.

CHAP. IX.

CHAP. X.

CHAP. XI.

NATURAL HISTORY.

CHAP. VI.

WEB-FOOTED WATERFOWL.

OF the web-footed waterfowl, the few which are distinguished by the name of *long-legged,* have so near an affinity with the birds of the preceding order, that some naturalists have classed them among the cranes, or waders; and, indeed, were it not for the very accurate distinction which the form of the foot affords, analogy would direct us to this arrangement in preference to every other.

THE AVOSET

Is easily distinguished from all other birds by the form of its bill, which is very thin, slender, and bends considerably upwards. The Scooping Avoset is about the size of the lapwing, or eighteen inches long; the bill is three inches and a half in length. The top of the head is black, the rest of the head, neck, and all the other parts of the body white, except the inner scapulars, the middle of the wing coverts and outer webs, and ends of the quills, which again are black. The legs are long, and of a bluish gray, and the toes have a connecting membrane. It weighs about thirteen ounces, and is frequent, in the winter, on most of the seacoasts of Europe, as well as in the fens of Lincolnshire, Cambridge, &c. in England. It feeds on worms and insects, which it scoops out of the sand with its bill. The American Avoset differs only in being something larger, and having the neck and breast of a deep cream colour. In Hudson's Bay there is a White Avoset.

AMERICAN AVOSET.

This species from its perpetual clammer and flippancy of tongue, is called by the inhabitants Cape May, the Lawyer. I found these birds, as well as the Long-legged Avoset, in the salt marshes of New-Jersey on the 20th of May. They flew around the shallow pools, uttering the sharp note of *click, click,* alighting on the marsh, or in the water, fluttering their loose wings, and shaking their half bent legs, as if ready to tumble over, keeping up a continual yelping note. The nest was built among the thick tufts of grass, of sea-weed, dry grass, and twigs, and raised to the height of several inches.—*Wilson.*

LONG LEGGED AVOSET.

This species arrives on the sea coast of New Jersey on the twenty fifth of April in small detached flocks of twenty or thirty together. They inhabit the marshes that are broken into numerous shallow pools, where they may be almost constantly seen wading often up to the breast in water. They feed on minute shell-fish, aquatic insects, and on eggs and spawn, that are deposited in the mud. This bird is known on the sea coast as the Stilt, Tilt, or Long Shanks; and naturalists have unaccountably classed it with the genus Plover.—*Wilson.*

THE COURIER

Is an Italian bird, somewhat less than the avoset, the bill is shorter, straight, and yellow. The upper parts of the plumage are of a rusty brown, the under parts white. It is remarkable for its swiftness in running, from which property it derives its name.

THE FLAMINGO

Is, perhaps, the most remarkable of waterfowl; it is one of the tallest, and the most beautiful. The body, which is of a beautiful scarlet, is no bigger than that of a swan; but its legs and neck are of such an extraordinary length, that when it stands erect, it is six feet six inches high. Its wings, extended, are five feet six inches from tip to tip; and it is four feet eight inches from tip to tail. The head is round and small, with a large bill, seven inches long, partly red, partly black, and crooked like a bow. The legs and thighs, which are not much thicker than a man's finger, are about two feet eight inches high; and its neck near three feet long. The feet are feeble, and united by membranes, as in those of the goose. Of what use these membranes are does not appear, as the bird is never seen swimming, its legs and thighs being sufficient to bear it into those depths where it seeks for prey.

This extraordinary bird is now chiefly found in America, but was once known on all the coasts of Europe. It is still occasionally met with on the shores of the Mediterranean. Its beauty, its size, and the peculiar delicacy of its flesh, have been such temptations to destroy or take it, that it has long since deserted the shores frequented by man, and taken refuge in countries that are as yet but thinly peopled.

When the Europeans first came to America, and coasted down along the African shores, they found the Flamingos on several shores on either continent gentle, and no way distrustful of mankind. When the fowler had killed one, the rest of the flock, far from attempting to fly, only regarded the fall of their companion in a kind of fixed astonishment: another and another shot was discharged; and thus the fowler often levelled the whole flock, before one of them began to think of escaping.

But at present it is very different in that part of the world; and the Flamingo is not only one of the scarcest, but one of the shyest birds in the world, and the most difficult of approach. They chiefly keep near the most deserted and inhospitable shores; near salt water lakes and swampy islands. When seen by mariners in the day, they always appear drawn up in a long close line of two or three hundred together; and, as Dampier tells us, present, at the distance of half a mile, the exact representation of a long brick wall. This line, however, is broken when they seek for food; but they always appoint one of the number as a watch, whose only employment is to observe and give notice of danger while the rest are feeding. As soon as this trusty sentinel perceives the remotest appearance of danger, he gives a loud scream, with a voice as shrill as a trumpet, and instantly the whole cohort are upon the wing. The flesh of the old ones is black and hard, though, Dampier says, well tasted: that of the young ones is better. But, of all delicacies, the Flamingo's tongue is the most celebrated. In fact, the Roman emperors considered them as the highest luxury; and we have an account of one of them, who procured fifteen hundred Flamingoes' tongues to be served up in a single dish. The tongue of this bird, which is so much sought after, is a good deal larger than that of any other bird whatever. The bill of the Flamingo is like a large black box, of an irregular figure, and filled with a tongue which is black and gristly.

Their time of breeding is according to the climate in which they reside: in North America they breed in summer; on the other side of the line they take the most favourable season of the year. They build their nests in extensive marshes, and where they are in no danger of a surprise. The nest is not less curious than the animal that builds it: it is raised from the surface of the pool about a foot and a half, formed of mud, scraped up together, and hardened by the sun, or the heat of the bird's body: it resembles a truncated cone, or one of the pots which we see placed on chimneys; on the top it is hollowed out to the shape of the bird, and in that cavity the female lays her eggs, without any lining but the well cemented mud that forms the sides of the building. She always lays two eggs, and no more; and, as her legs are immoderately long, she straddles on the nest, while her legs hang down, one on each side, into the water. The young ones are a long while before they are able to fly; but they run with amazing swiftness. They are sometimes caught; and, very different from the old ones, suffer themselves to be carried home, and are tamed very easily.*

THE AUK.

THE first European bird of the web-footed fowls with short legs, which naturalists introduce to our notice, is the Auk, of which there are about twelve species. The whole tribe is distinguished peculiarly by the form of the bill, which is strong, convex, compressed at the sides, in general crossed with several furrows, and in some degree resembling the coulter of a plough.

THE GREAT AUK

Is the size of a goose; its bill is black, about four inches and a quarter in length, and covered at the base with short velvet-like feathers. The upper parts of the plumage are black, and the lower parts white, with a spot of

* This singular species are seen on the southern frontiers of the U. States, and on the peninsula of East Florida. They are also common in the warmer parts of America, Peru, Chili, Cayenne, Brazil and the West Indies.—*Wilson.*

white between the bill and the eyes, and an oblong stripe
of the same on the wings, which are too short for flight.
The bird is also a very bad walker, but swims and dives
well. It is, however, observed by seamen, that it is never
seen out of soundings, so that its appearance serves as an
infallible direction to land. It feeds on the lump fish and
others of the same size; and is frequent on the coasts of
Norway, Greenland, Newfoundland, &c. It lays its eggs
close to the seamark.

THE RAZORBILL

Is not above half the size of the preceding, which it resembles both in form and plumage, except that it has the use
of its wings, and lays its egg (for each of these species
lays but one) on the bare top of a precipice, and fastens it
by a cement, so as to prevent its rolling off. It is pretty
common on the coasts of England during the summer season. The BLACKBILLED AUK is still smaller.

THE PUFFIN

Is the size of the teal, weighs about twelve ounces, and is
twelve inches in length. The bill is much compressed;

the half next the point is red, that next the base is blue gray. It has three furrows or grooves impressed in it; one in the livid part, two in the red. The eyes are fenced with a protuberant skin, of a livid colour; and they are gray or ash coloured.

The Puffin, like all the rest of this kind, has its legs thrown so far back, that it can hardly move without tumbling. This makes it rise with difficulty, and subject to many falls before it gets upon the wing; but as it is a small bird, when it once rises, it can continue its flight with great celerity

All the winter these birds are absent, visiting regions too remote for discovery. At the latter end of March, or the beginning of April, a troop of their spies or harbingers come and stay two or three days, as it were to view and search out their former situations, and see whether all be well. This done, they once more depart; and, about the beginning of May, retnrn again with the whole army of their companions. But if the season happens to be stormy and tempestuous, and the sea troubled, the unfortunate voyagers undergo incredible hardships; and they are found, by hundreds, cast away upon the shores, lean and perished with famine.

The Puffin, when it prepares for breeding, which always happens a few days after its arrival, begins to scrape out a hole in the ground, not far from the shore; and when it has penetrated some way into the earth, it then throws itself upon its back, and with its bill and claws thus burrows inward, till it has dug a hole with several windings and turnings, from eight to ten feet deep. It particularly seeks to dig under a stone, where it expects the greatest security. In this fortified retreat it lays one egg; which, though the bird be not much bigger than a pigeon, is the size of a hen's.

Few birds or beasts will venture to attack them in their retreats. When the great sea raven comes to take away their young, the Puffin boldly opposes him. Their meeting affords a most singular combat. As soon as the raven approaches, the Puffin catches him under the throat with its beak, and sticks its claws into its breast, which makes the raven, with a loud screaming, attempt to get away; but the little bird still holds fast to the invader, nor lets him go till they both come to the sea, where they drop down together, and the raven is drowned: yet the raven is but too often successful; and, invading the Puffin at the bottom of its hole, devours both the parent and its family.

The Little Auk is still less than the puffin, being not above the size of a blackbird.

Of the birds of this genus, the Tufted Auk is one of the most curious. It is somewhat bigger than the puffin, and is distinguished by a tuft of feathers, four inches in length, which arises over each eye, and falls elegantly on each side of the neck. It is found at Kamtschatka.

The Crested Auk is perhaps still more remarkable, having its head adorned with a crest, composed of long feathers, and which curves forward over the bill. This bird inhabits the islands contiguous to Japan. Besides these, there are the Parroquet and Dusky Auk, and some other species of less note.

THE GUILLEMOT

Is nearly allied to the preceding genus, but it wants the characteristic bill, which in this genus is slender, strong, and pointed.

The largest species with which we are acquainted is the Foolish Guillemot, which weighs about twenty-four ounces, and is seventeen inches in length. The bill is black, and three inches in length. The head, neck, back, wings, and tail, are of a deep mouse colour; the tips of the lesser quill feathers, and all the under parts of the plumage are white. They accompany the auk in its visits

to the northern shores, and are such foolish birds that they will not quit the rock, though they see their companions killed around them.

The LESSER GUILLEMOT weighs about fifteen ounces. The upper parts of its plumage are darker than in the former species. The BLACK GUILLEMOT is entirely black, except a large mark of white on the wings. In winter, however, this bird is said to change to white; and there is a variety in Scotland not uncommon, which is spotted, and which Mr. Edwards has described under the name of the spotted Greenland Dove. The MARBLED GUILLEMOT, which is found at Kamtschatka, &c. receives its name from its plumage, which is dusky, elegantly marbled with white.

THE DIVER GENUS

INCLUDES about seven species. The great NORTHERN DIVER, or LOON,* weighs sixteen pounds, and measures

---

* This bird is found along the American shores, and are said to win-

three feet six inches in length. The bill is strong, black, and above four inches in length. The head and neck are vel-vet black, with a white crescent immediately under the' throat, and another behind. The upper parts of the plumage are also black, spotted with white, and the breast and belly perfectly white. This bird is found in all the northern parts of Europe, and feeds on fish. It breeds on the inaccessible rocks and steep cliffs in the Isle of Man, and likewise in Cornwall; on Prestholm Island, near Beaumaris in the Isle of Anglesey; also on the Fern Island, near Northumberland, and in the cliffs about Scarborough, in Yorkshire; and several other places in England. The Northern Diver lays exceedingly large eggs; being full three inches long, blunt at one end, and sharp at the other, of a sort of bluish colour, generally spotted with some black spots or strokes. It flies high and well.

The IMBER is less than the preceding, but' still larger than a goose. The upper parts of the plumage are in general dusky; the under parts silvery white. It is very common in the Orkneys. The skins of both these species are so remarkably tough, that in some of the northern countries they have been used as leather.

The SPECKLED DIVER is more common in the southern parts of Europe. It is called on the Thames the *Sprat Loon*. It weighs about two pounds and a half; and has the upper parts of the body dusky spotted with white, the breast and belly white. It is so confident of its skill in diving, that it often approaches very near the boats when fishing for sprats or herrings.

The BLACK-THROATED DIVER has the fore part of the

ter at Chesapeake Bay. In the summer, they retire to the fresh water lakes, and they are reported to breed in the New England States. In winter, they migrate to the southern States.—*Wilson.*

throat black, the back and wings of the same colour spotted with white; the head and neck ash colour, and the breast and belly white. This bird is common in the northern regions, but seldom found in France or England.

The RED-THROATED DIVER differs chiefly from the preceding in having the throat and part of the neck of a fine red; the upper parts of the body also are dusky, marked with a few white spots. It is seldom found to the south of Scotland.

Of the foreign birds of this species, the CHINESE DIVER is the only animal worth notice. The upper parts of the plumage are of a greenish brown; the under parts a reddish white, marked with dark spots. This is generally supposed to be one of the birds which the Chinese train up for the purpose of fishing, of which we shall have occasion to treat further when we speak of the corvorant.

THE TERN.*

OF the Tern there are about twenty-three different species, which are all distinguished by one common characteristic, viz. the forked tail.

THE GREAT TERN

Is about fourteen inches long, and weighs four ounces and a quarter. The bill and feet are a fine crimson: the former is tipped with black, and very slender. The back of the head is black; the upper part of the body a pale gray,

---

* Turton enumerates twenty-five species, six of which are natives of the United States. They are known to the people residing on the sea shore, by the name of Sea Swallows, and some of them are confounded with the Gulls. Our Great Tern is the same as that of Europe. The Lesser Tern is also the same in both countries.—*Wilson.*

and the under part white. These birds have been called sea swallows, as they appear to have all the same actions at sea that the swallow has at land, seizing every insect

which appears on the surface, and darting down upon the smaller fishes, which they seize with incredible rapidity.

The LESSER TERN weighs only two ounces and five grains. The bill is yellow; and from the eyes to the bill is a black line. In other respects it almost exactly resembles the preceding.

The BLACK TERN is of a middle size between the two preceding species. It weighs two ounces and a half. It receives its name from being all black as far as the vent, except a spot of white under the throat. This bird is called, about Cambridge in England, the *Car Swallow*. It is a very noisy animal.

Among the foreign birds of the Tern genus, there are some found of a snowy white; but the most singular bird of this kind is the STRIATED TERN which is found at New Zealand. It is thirteen inches in length. The bill is black, and the body in general mottled, or rather striped, with black and white. The NODDY * is about fifteen inch-

* This bird has been long known to navigators, as its appearance at

es long. The bill is black, and two inches long, and the whole plumage a sooty brown, except the top of the head, which is white. It is a very common bird in the tropical seas, where it is known frequently to fly on board ships, and is taken with the hand. But though it be thus stupid, it bites the fingers severely, so as to make it unsafe to hold it. It is said to breed in the Bahama Islands.

MARSH TERN.

THIS new species I met with on the shores of Cape May, particularly over the salt marshes, and darting down after a kind of large black spider, common in such places. This spider can travel under water as well as above, and during summer, at least, seems to constitute the food of this Tern.—*Wilson.*

SHORT TAILED TERN.

ON the sixth of September, 1812, after a violent northeast storm which inundated the meadows of Schuylkill in many places, numerous flocks of this Tern all at once made their appearance, flying over the watery spaces, picking up grasshoppers, spiders, beetles, and other insects that were floating on the surface. The people on the sea coast say that this bird comes to them only in the fall, and is frequently seen about mill ponds and fresh water marshes.
Wilson.

THE PETRELS.

THE whole genus of Petrels are known by having, instead of a back toe, only a sharp spur or nail; they have also a
sea usually indicates the vicinity of land. It is widely dispersed over the various shores of the ocean. I observed them on the coast of Florida and Georgia, where they were very numerous and noisy. They frequently settle on the rigging of ships at sea, and are called by the sailors, the Noddy.—*Wilson.*

faculty of spouting from their bills, to a considerable distance, a large quantity of pure oil, which they do, by way of defence, into the face of any person who attempts to take them.

THE FULMAR

Is the largest of the kind which is known in Europe. It is superior to the size of the common gull, being about fifteen inches in length, and in weight seventeen ounces. The bill is very strong, yellow, and hooked at the end. The head, neck, and all the under parts of the body, are white ; the back and wings ash-coloured, the quills dusky, and the tail white. It feeds on the blubber of whales which supplies the reservoir, whence it spouts, with a constant stock of ammunition. This oil is esteemed by the inhabitants of the North as a sovereign remedy in many complaints both external and internal. The flesh is also considered by them as a delicacy, and the bird is therefore in great request at St. Kilda. When a whale is taken, these birds will, in defiance of all endeavours, light upon it, and pick out large lumps of fat, even while it is alive.

The SHEARWATER, or MANKS PUFFIN, as it is called by Willoughby, is something smaller than the preceding.

The head and all the upper part of the body are of a sooty blackness; and the under part and inner coverts of the wings white. These birds are found in the Isle of Man and the Scilly Isles. In February they take a short possession of the rabbit burrows, and then disappear till April; they lay one egg, and in a short time the young are fit to be taken. They are then salted and barrelled. During the day they keep at sea fishing, and towards evening return to their young, whom they feed by discharging the contents of the stomach into their mouths.

THE STORMY PETREL *

Is about the size of a house swallow. The general colour of the plumage is black, except about the rump, which is white. Stormy Petrels have been seen in flocks which have been estimated to contain at least a hundred and fifty millions of them. They are always to be found on the shores of Britain, and seem to be diffused all over the world. They sometimes hover over the water like swallows, and sometimes appear to run on the top of it: they are also excellent divers. It skims along the hollows of

* Wilson supposed the American Stormy Petrel to be the same as that of Europe, but Charles Bonaparte has shown that it is a distinct species. It breeds in great numbers on the shores of the Bahama and Bermuda Isles, and on the Coast of East Florida and Cuba. This author enumerates four species of the Stormy Petrel.

the waves, and through the spray upon their tops, at the astonishing rate of sixty miles in an hour. They are very clamorous, and are called by the sailors *Mother Cary's Chickens*, who observe they never settle or sit upon the water but when stormy weather is to be expected. They are found in most parts of the world; and in the Feroe islands the inhabitants draw a wick through the body of the bird, from the mouth to the rump, which serves them as a candle, being fed by the vast proportion of oil which this little animal contains. This oil it is supposed to collect from the ocean by means of the feathers on its breast.

There are about twenty species of foreign birds of this kind. In the high southern latitudes one is found which is the size of a goose, and on that account called the GIANT PETREL. The upper parts of its plumage are pale brown, mottled with dusky white; the under parts are white. There is another species in Norfolk Island, which burrows in the sand like a rabbit.

THE GULL,

AND all its varieties, is well known to most readers. It is seen with slow-sailing flight hovering over rivers, to prey upon the smaller kinds of fish: it is seen following the ploughman in fallow fields to pick up insects; and when living animal food is not to be found, it has even been known to eat carrion, and whatever else offers of the kind.

Of the Gull there are about nineteen species. The largest with which we are acquainted is the BLACK AND WHITE or BLACK-BACKED GULL. It generally weighs upwards of four pounds, and is twenty-five or twenty-six inches from the point of the bill to the end of the tail; and from the tip of each wing, when extended, five feet and several inches. The bill appears compressed sideways, being more than three inches long, and hooked towards the end, like the rest of this kind, of a sort of orange colour; the nostrils are of an oblong form; the mouth is wide, with a long tongue and very open gullet. The irides of the eyes are of a delightful red. The wings and the middle of the back are black; only the tips of the covert and quill feathers are white. The head, breast, tail, and other parts of the body, are likewise white. The tail is near six inches long, the legs and feet are flesh-coloured, and the claws black. There are about twenty varieties of this tribe, which are all distinguished by an angular knob on the chap.

Gulls are found in great plenty in every place; but it is chiefly round the rockiest shores that they are seen in the greatest abundance: it is there that the Gull breeds and brings up its young; it is there that millions of them are heard screaming with discordant notes for months together.

The SKUA GULL is the size of a raven. The upper parts of the head, neck, back, and wings, are deep brown; the under parts a pale rusty ash colour. The legs are black, rough, and warty, and the talons very strong and hooked. It is mostly a native of the North, though often found in England. It is a most formidable bird, as it not only preys upon fish, but upon all the smaller waterfowl, and even on young lambs. It has the fierceness of the eagle in defending its young; and when the inhabitants of the Faro isles attack its nest, they hold a knife over their heads, on which the Skua will transfix itself in its fall on the invaders. On the rocky island of Foula, one of the Shetland isles, it is a privileged bird, as it is said to defend the flocks from the eagle, which it pursues and beats off with great fury whenever he presumes to visit the island.

The WAGEL GULL has its whole plumage composed of a mixed brown ash colour and white. It weighs about three pounds.

The HERRING GULL resembles the black-backed in every thing but size, and that the plumage on the back and wings is more inclined to ash colour than black; it weighs thirty ounces. The GLAUCOUS GULL, or BURGOMASTER, which inhabits Norway, &c. is rather larger than the Herring Gull, but resembles it in most other respects. The SILVERY GULL is the same size as the Herring Gull, and not much different in plumage and manners.

The TARRACK and the KITTIWAKE GULLS also so nearly resemble each other, that some authors affirm the latter to be only the Tarrack in a state of perfection. The head, neck, belly, and tail of the Kittiwake are of a snowy whiteness; the back and wings are gray; and both species have behind each ear a dark spot: both species are about the same size, viz. fourteen inches; and the Tarrack weighs seven ounces. Of the ARCTIC GULL the male has the top of the head black; the back, wings, and tail dusky; the rest of the body white: the female is entirely brown. It has been called the parasite, from its habits of pursuing the lesser Gulls till they drop their ordure through fear, which this filthy animal catches and devours before it drops into the water.

The COMMON GULL is seventeen inches long, and weighs one pound. The bill is yellow; the back and wings a pale

gray; and the head and rest of the body white. The
WINTER GULL is also very common in all these parts of
Europe. The top of the head is white, marked with oblong dusky spots; the back and wings ash colour, marked
with dusky brown.

The jelly-like substance known by the name of star-shot
or star-jelly, owes its origin to some of these birds, being
nothing but the half-digested remains of earthworms, on
which they feed, and often discharge from their stomachs.

The PEWIT GULL, or BLACK-CAP, is so called from the
head and throat being of a dark or black colour. The RED
LEGGED GULL, the BROWN-HEADED GULL, the LAUGHING
GULL,* which only differs from the others in having the
legs black instead of red, are possibly only varieties of the
same species. They are in length from fifteen to eighteen
inches. The back and wings of these birds are in general
ash colour, and the rest of the body white. The young
birds of these species are thought by some to be good eating.

The GOAT GULL, which is found on the borders of the
Caspian Sea, though distinguished by a black head, is quite
a different species from the black-cap, as it equals in size
the Barnacle goose, and weighs between two and three
pounds: its voice too is as hoarse as that of a raven.

The Gull genus, like all other rapacious birds, lay but
few eggs; and hence, in many places, their number is daily
seen to diminish. Most of the kind are fishy tasted, with
black stringy flesh; and of these, the poor inhabitants of
the northern British islands make their wretched banquets.
They have been long used to no other food; and even salted
Gull can be relished by those who know no better.

---

* This bird, called the Black-headed Gull in America, appears in
New Jersey in the latter part of April. They breed in marshes, and
live on worms, insects and animal substances.—*Wilson.*

The Gull, the petrel, the tern, have all nearly the same habits, the same nature, and are caught in the same manner; that is, at the most imminent risk, and with the loss of many lives in the course of a season.

But of this dangerous sport a more particular description will perhaps be acceptable to the reader. Those who have been upon the British coasts know that there are two different kinds of shores; that which slopes towards the water with a gentle declivity, and that which rises with a precipitate boldness, and appears as a bulwark to repel the force of the invading deep. It is to such shores as these that the vast variety of seafowl resort, and in the cavities of these rocks they breed in safety. Of the tremendous sublimity of these elevations it is not easy to form an idea. The boasted works of art, the highest towers, the noblest domes, are but ant-hills, when put in comparison; the single cavity of a rock often exhibits a coping higher than the ceiling of a Gothic cathedral. What should we think of a precipice three quarters of a mile in height? and yet the rocks of St. Kilda are still higher! What must be our awe to approach the edge of that impending height, and to look down on the unfathomable vacuity below! To ponder on the terrors of falling to the bottom, where the waves, that swell like mountains, are scarcely seen to curl on the surface, and the roar of the ocean appears softer than the murmur of a brook! It is in these formidable mansions that myriads of seafowls are ever seen sporting. To the spectator from above, those birds, though some of them above the size of an eagle, seem scarcely as large as a swallow: and their loudest screaming is scarce perceptible.

Yet even here these animals are not in perfect security from the arts and activity of man. Want, which is the great spring of human exertion, can force the cottager to tempt the most formidable dangers, and to put forth an endeavour almost beyond the force of man. When the precipice is to be assailed from below, the fowlers furnish themselves with poles of five or six ells in length, with a hook at the end, and fixing one of these poles in the girdle of the person who is to ascend, his companions, in a boat, 'or on a projection of the cliff, assist his progress till he procures a firm footing. When this is accomplished, he draws the others up with a rope, and another man is forwarded again by means of the pole to a higher station. Frequently the person who is in the highest situation holds another man suspended by a rope, and directs his course to the place where the birds have placed their nests. It unfortunately too often happens that the man who holds the rope has not a footing sufficiently secure, and in that case both of them inevitably perish.

Some precipices are so abrupt, that they are not by any means to be ascended from below. In this case a rope is provided of eighty or a hundred fathoms long, which one of the fowlers fastens to his waist, and between his legs, in such a manner as to support him in a sitting posture. The rope is held by five or six persons on the top, and it slides upon a piece of wood, which is laid so as to project beyond the precipice. By means of this apparatus, the man is gradually let down, and he attacks the habitations of the feathered race with the most sanguinary success. This operation is, however, not without its dangers. By the descent and friction of the rope the loose stones are furiously hurled down on every side. To defend himself from their blows, the fowler covers his head with a kind of helmet, or with a seaman's shaggy cap. Many, however, lose their lives by this kind of accident. Those who are unskilful, frequently suffer by a giddiness with which they are seized, on beholding themselves suspended from this tremendous height: he, on the other hand, who is accustomed to the sport, swings himself about with amazing dexterity ; he directs his attack to that part of the rock which promises the amplest success ; with his fowling staff he strikes the game as it proceeds out of the holes ; he occasionally disengages himself from the rope by which he was suspended ; he roams through the cavities of the rock, and when he has provided himself with a sufficient booty, he gives the signal to his companions, and is again drawn up, when the festivity of the evening, among these poor and desperate adventurers, generally compensates for the fatigues and dangers of the day.

THE MERGANSER GENUS

INCLUDES only about six species, in all of which the bill is slender, and furnished at the end with a crooked nail, and grated or toothed both upper and under chap like a saw. The largest of the kind is the GOOSEANDER,* which weighs about four pounds. The bill is red; the head very full of feathers on the top and back part. The plumage is various and beautiful. The head and upper parts are fine glossy black, the rump and tail ash colour, and the under parts of the neck and body a fine pale yellow. Its manners and appetites entirely resemble those of the diver. It feeds upon fish, for which it dives; it is said to build its nest upon trees, like the heron and the corvorant.

The DUN DIVER is less than the gooseander. The upper part of the head is reddish brown ; the back and wings ash colour; and the lower parts of the body white. It is

* This bird is a winter inhabitant only, of the sea shores, fresh water lakes and rivers of the United Sates. It comes in November, and disappears in April. We have no account of their manner of breeding.

found in the same places and has the same manners with the gooseander. The RED-BREASTED MERGANSER is still smaller, weighing only two pounds. The head and neck are black, glossed with green; the rest of the neck and the belly white: the upper part of the back is glossy black; the lower parts and the rump are striated with brown and pale gray: on the wings there are white bars tipped with black, and the breast is reddish, mixed with black and white. The plumage of the female is less splendid; and they differ in another respect, viz. that the male has a very full and large crest, the female only the rudiment of one. It is common on the shores of the United States as well as in Europe.

THE SMEW, OR WHITE-HEADED GOOSEANDER,

MEASURES eighteen inches from the point of the bill to the extremity. It has a fine crest upon the head, which falls down towards the back part of it, under which, on each side of the head, is a black spot: the rest of the head and neck, and the under parts of the body, are white; the back and the wings are a pleasing mixture of black and white. The tail is about three inches long, of a kind of dusky ash colour, the feathers on each side shortening gradually.

The female has no crest ; the sides of the head red; the wings of a dusky ash colour; the throat is white. In other respects it agrees with the male.

The MINUTE MERGANSER is still less than the smew. The head is slightly crested, and of a rust colour; the back and tail are of a dusky ash colour; the breast mottled, and the belly white.

The HOODED MERGANSER is a native of North America, and peculiar to that country. It is common on the coasts of New England, and breeds in the arctic regions. It is about the size of a widgeon. The head and neck are dark brown, the former surrounded with a large round crest, the middle of which is white. The back and quills are black, the tail dusky ; and the breast and belly white, undulated with black. The female is fainter in the colour of her plumage, and has a smaller crest.

THE DUCK GENUS

EMBRACES one hundred species, infinitely differing in size and plumage ; many of them are rendered domestic, but a still greater proportion are in their native untamed state. All the species are distinguished by their strong flat bill, furnished at the end with an additional piece, termed a nail, and marked at the edges with lamellæ, or teeth.

Though these birds do not reject animal food when offered them, yet they can contentedly subsist upon vegetables, and seldom seek any other. They are easily provided for; wherever there is water there seems to be plenty. All the other web-footed tribes are continually voracious, continually preying. These lead more harmless lives: the weeds on the surface of the water, or the insects at the bottom, the grass by the bank, or the fruits and corn in cultivated grounds, are sufficient to satisfy their easy appetites.

They breed in great abundance, and lead their young to the pool the instant they are excluded.

As their food is simple, so their flesh is nourishing and wholesome. The swan was considered as a high delicacy among the ancients; the goose was abstained from as totally indigestible. Modern manners have inverted tastes; the goose is now become the favourite; and the swan is seldom brought to table, unless for the purpose of ostentation. But at all times the flesh of the duck was in high esteem; the ancients thought even more highly of it than we do. We are contented to eat it as a delicacy; they also considered it as a medicine: and Plutarch assures us, that Cato kept his whole family in health, by feeding them with duck whenever they threatened to be out of order.

THE SWAN.

So much difference is there between this bird when on land and in the water, that it is hardly to be supposed the same, for in the latter, no bird can possibly exceed it for beauty and majestic appearance. When it ascends from its favourite element, its motions are awkward, and its neck is stretched forward with an air of stupidity; it has, indeed, the air of being only a larger sort of goose; but when seen smoothly gliding along the water, displaying a thousand graceful attitudes, and moving at pleasure without the smallest apparent effort, there is not a more beautiful figure in all nature. In its form, we find no broken or harsh lines; in its motions, nothing constrained or abrupt, but the roundest contours, and the easiest transitions; the eye wanders over the whole with unalloyed pleasure, and with every change of position every part assumes a new grace. It will swim faster than a man can walk.

This bird has long been rendered domestic; and it is now a doubt whether there be any of the tame kind in a state of nature. The colour of the tame Swan is entirely white, and it generally weighs full twenty pounds. Under the feathers is a very thick soft down, which is made an article of commerce, for purposes of both use and ornament. The windpipe sinks down into the lungs in the ordinary manner; and it is the most silent of all the feathered tribe; it can do nothing more than hiss, which it does on receiving any provocation. In these respects it is very different from the wild or whistling Swan.

The beautiful bird is as delicate in its appetites as it is elegant in its form. Its chief food is corn, bread, herbs growing in the water, and roots and seeds, which are found near the margin. At the time of incubation it prepares a nest in some retired part of the bank, and chiefly where there is an islet in the stream. This is composed of water plants, long grass, and sticks: and the male and female assist in forming it with great assiduity. The Swan lays seven or eight white eggs, one every other day, much larger than those of a goose, with a hard, and sometimes a tuberous shell. It sits six weeks before its young are excluded; which are ash coloured when they first leave the shell, and for some months after. It is not a little dangerous to approach the old ones, when their little family are feeding around them. Their fears as well as their pride seem to take the alarm, and when in danger, the old birds carry off the young ones on their back. A female has been known to attack and drown a fox, which was swimming towards her nest: they are able to throw down and trample on youths of fifteen or sixteen; and an old Swan can break the leg of a man with a single stroke of its wing.

Swans were formerly held in such great esteem in England, that, by an act of Edward the Fourth, none, except the son of the king, was permitted to keep a Swan, unless possessed of a freehold to the value of five marks a year. By a subsequent act, the punishment for taking their eggs was imprisonment for a year and a day, and a fine at the king's will. At present they are not valued for the delicacy of their flesh; but numbers are still preserved for their beauty. Many may be seen on the Thames, where they are esteemed royal property, and it is accounted felony to steal their eggs. On this river, as far as the conservancy of it belongs to the city of London, they are under the care of the corporation; and at certain times the lord mayor, aldermen, &c. proceed up the Thames, to what is commonly called the Swan hopping, to mark the young birds. The Swan is a long-lived bird, and sometimes attains the age of more than a hundred years.

The WILD or WHISTLING SWAN, though so strongly resembling this in colour and form, is yet a different bird; for it is very differently formed within. The wild Swan is less than the tame, almost a fourth; for as the one

weighs twenty pounds, the other only weighs sixteen
pounds and three quarters. The colour of the tame Swan
is all over white; that of the wild bird is along the back
and the tips of the wings of ash colour; the tame Swan is

mute, the wild one has a sharp loud cry, particularly while
flying. But these are slight differences, compared to what
are found upon dissection. The wild species is found in
most of the northern regions, in America, and probably in
the East Indies.

The BLACK SWAN. New Holland, that country of animal wonders, presents us with a bird which the ancients
imagined could not possibly have existence. The Black
Swan is exactly similar in its form to the Swan of the old
world, but is somewhat smaller in size. Every part of its
plumage is perfectly black, with the exception of the pri-
mary and a few of the secondary quill feathers, which are

white. The bill is of a bright red above, is crossed at the anterior part by a whitish band; is of a grayish white on the under part; and, in the male, is surmounted at the base by a slight protuberance. The legs and feet are of a

dull ash colour. Black Swans, in their wild state, are extremely shy. They are found in Van Dieman's Land, New South Wales, and on the western coast of New Holland; and are generally seen swimming on a lake, in flocks consisting of eight or nine individuals. On being disturbed, they fly off in a direct line one after the other, like wild geese.

THE GOOSE,

In its domestic state, exhibits a variety of colours. The WILD GOOSE, or GRAY LAG, always retains the same marks: the whole upper part is ash coloured; the breast and belly are of a dirty white; the bill is narrow at the base, and at the tip it is black; the legs are of a saffron colour, and the claws black. It frequently weighs about ten pounds.

The Wild Goose is supposed to breed in the northern parts of Europe; and, in the beginning of winter, to descend into more temperate regions. If they come to the ground by day, they range themselves in a line, like cranes; and seem rather to have descended for rest, than for other refreshment. When they have sat in this manner for an hour or two, we have heard one of them, with a loud long note, sound a kind of charge, to which the rest punctually attended, and they pursued their journey with renewed alacrity. Their flight is very regularly arranged; they either go in a line abreast, or in two lines, joining in an angle in the middle.

The common TAME GOOSE is nothing more than the Wild Goose in a state of domestication. It is sometimes found white, though much more frequently verging to gray; and it is a dispute among men of taste, which should have the preference.

These birds, in rural economy, are an object of attention and profit, and are no where kept in such vast quantities as in the fens of Lincolnshire in England; several persons there having as many as a thousand breeders. They are bred for the sake of their quills and feathers; for which they are stripped while alive, once in the year for their quills, and no less than five times for the feathers: the first plucking commences about Lady Day, for both; and the other four between Lady Day and Michaelmas. It is said that in general the birds do not suffer much from this operation; except cold weather sets in, which then kills great numbers of them. The old Geese submit quietly to be plucked, but the young ones are very noisy and unruly. The possessors, except in this cruel practice, treat their birds with kindness, lodging them very often even in the same room with themselves.

These Geese breed in general only once a year, but if well kept they sometimes hatch twice in a season. During their sitting, each bird has a space allotted to it, in rows of wicker pens placed one above another; and it is said that the gozzard, or gooseherd, who has the care of them, drives the whole flock to water twice a day, and bringing them back to their habitations, places every bird in its own nest. The numbers of Geese which are sent to London for sale, are enormous, two or three thousand being frequently seen in a drove, and some droves having even contained more than nine thousand.

The BEAN GOOSE is a bird of passage, and arrives in Lincolnshire about autumn, and departs in May. It weighs about six pounds. The bill is smaller than in the preceding species. The head and neck are brown, the back and wings ash colour, and the breast and belly dirty white. It feeds much on the young corn, beans, &c. whence its name.

The BARNACLE differs in some respects from all these; being less than any of them, with a black bill, much shorter than any of the preceding. It is scarcely necessary to combat the idle error of this bird's being bred from the shell sticking to ships' bottoms; it is well known to be hatched from an egg, in the ordinary manner, and to differ in very few particulars from all the rest of its kind. The upper parts of the plumage are black, the forehead, chin, and all the under parts white.

The BRENT GOOSE is still less than the former, and not bigger than a Muscovy duck, except that the body is longer. The head, neck, and upper part of the breast, are black; about the middle of the neck, on each side, are two small spots, or lines of white, which together appear like a ring. Both this and the preceding frequent the coasts of England in winter; and in some seasons have been so numerous, on the coasts of Piccardy, as to become a pest to the inhabitants.

CANADA GOOSE.

THIS is the common Field Goose of the United States, universally known over the whole country; whose regular periodical migrations are the sure signals of returning spring, or approaching winter. I have never yet visited any quarter of the country, where the inhabitants are not familiarly acquainted with the passing and repassing of the Wild Geese. The general opinion here is, that they are on their way to the lakes to breed; but the inhabitants on the confines of the great lakes are equally ignorant with ourselves of the particular breeding places of those birds. *There*, their journey north is but commencing, and how far it extends it is impossible for us at present to ascertain. They were seen by Hearne in large flocks within the arctic circle, and were then pursuing their way still farther north. They have been seen also on the dreary coast of Spitsbergen, feeding on the water's edge.

It is highly probable that they extend their migrations under the very pole itself, amid the silent desolation of unknown countries, shut out from the eye of man by everlasting barriers of ice. That such places abound with suitable food we cannot for a moment doubt.

The flight of the Wild Geese is heavy and laborious, generally in a straight line or in two lines approximating to a point. In both cases, the van is led by an old gander, who every now and then pipes his well known *houk*, as if to ask how they come on; and the houk of " all 's well " is generally returned by some of the party. When bewildered in foggy weather, they appear sometimes to be in great distress, flying about in an irregular manner, making a great clamour. On these occasions, should they alight on the earth, as they sometimes do, they meet with speedy death and destruction. The autumnal flight lasts from the middle of August to the middle of October; the vernal flight from the middle of April to the middle of May.

Wounded Geese have frequently been domesticated, and readily pair with tame geese. On the approach of spring, however, they discover symptoms of great uneasiness, frequently looking up in the air and attempting to go off. Some, whose wings have been closely cut, have travelled on foot in a northerly direction, and have been found at the distance of several miles from home. They hail every flock that passes overhead, and the salute is sure to be returned by the voyagers, who are only prevented from alighting among them, by the presence and habitations of man. The gunners sometimes take one or two of these domesticated Geese with them to those places over which the wild ones are accustomed to fly; and concealing themselves, wait for a flight, which is no sooner perceived by the decoy Geese, than they begin calling aloud, until the flock approaches so near, that the gunners are enabled to make great havoc among them with their musket shot.

The English at Hudson's Bay depend greatly on Geese, and in favourable seasons kill three or four thousand, and barrel them up for use. They send out their servants as well as Indians to kill them on their passage. They mimic the cackle of the Geese so well, that many of them are allured to the spot where they are concealed, and are thus easily shot. When in good order, the Wild Goose weighs from ten to fourteen pounds, and is estimated to yield half a pound of feathers. It is domesticated in numerous quarters of the country, and is remarked for being extremely watchful, and more sensible of approaching changes in the atmosphere than the common Gray Goose. In England, France, and Germany, they have been long ago domesticated.

Mr. Platt, a respectable farmer on Long Island, being out shooting in one of the bays which in that part of the country abound in water-fowl, wounded a Wild Goose. Being unable to fly, he caught it, and brought it home alive. It proved to be a female, and turning it into his yard with a flock of tame Geese it soon became quite familiar, and in a little time its wounded wing entirely healed. In the following spring, when the Wild Geese migrate to the northward, a flock passed over Mr. Platt's barn-yard, and just at that moment, their leader, happening to sound his bugle note, our Goose, in whom its new habits and enjoyments had not quite extinguished the love of liberty, and remembering the well-known sound, spread its wings, mounted into the air, joined the travellers, and soon disappeared. In the succeeding autumn, the Wild Geese, as usual, returned from the northward, in great numbers, to pass the winter in our bays and rivers. Mr. Platt happened to be standing in his yard, when a flock passed directly over his barn. At that instant, he observed three Geese detach themselves from the rest, and after wheeling round several times, alight in the middle of the yard. Imagine his surprise and pleasure, when, by certain well remembered signs, he recognised in one of the three his long lost fugitive. It was she indeed! She had travelled many hundred miles to the lakes; had there hatched and reared her offspring; and had now returned with her little family, to share with them the sweets of civilized life. The birds were all living, and in Mr. Platt's possession, a year ago, and had shown no disposition whatever to leave him.—*Wilson.*

THE TAME DUCK

Is the most easily reared of all our domestic animals. The WILD DUCK, or MALLARD, differs, in many respects, from the tame; and in them there is a still greater variety than among the domestic kinds.

The most obvious distinction between wild and tame Ducks is in the colour of their feet; those of the tame Duck being black; those of the wild Duck yellow. The difference between wild Ducks among each other, arises as well from their size, as the nature of the place they feed in. Sea Ducks, which feed in the salt water, and dive much, have a broad bill, bending upwards, a large hind toe, and a long blunt tail. Pond Ducks, which feed in plashes, have a straight and narrow bill, a small hind toe, and a sharp-pointed train. The former are called in England by the decoy-men foreign Ducks; the latter are supposed to be natives of England. In this tribe, we may rank, as natives of Europe, the EIDER DUCK, which is double the size of a common Duck, with a black bill; the VELVET DUCK, not so large, and with a yellow bill; the SCOTER DUCK, or BLACK DIVER, with a knob at the base of a yellow bill; the TUFTED DUCK, adorned with a thick crest; the SCAUP DUCK, less than the common Duck, with the bill of a grayish blue colour; the GOLDEN EYE, with a large white spot at the corners of the mouth, resembling an eye; the SHELDRAKE, with the bill of a bright red, and swelling into a knob; the MALLARD, which is the stock whence the tame breed has probably been produced; the SHOVELLER, which has a bill three inches long, and remarkably broad at the end; the PINTAIL, with the two middle feathers of the tail three inches longer than the rest; the POCHARD, with the head and neck of a bright bay; the LONG-TAILED Duck, the general colour of whose plumage is deep chocolate, and the outer feathers of the tail, which are white, four inches longer than the rest; the WIDGEON, and lastly, the TEAL. Of these we shall describe more particularly the Eider Duck, the Widgeon, and the Teal.

THE EIDER DUCK

HAS a black cylindrical bill, and the feathers of the forehead and cheeks advance far into the base. In the male, the feathers of part of the head, the lower part of the breast, the belly, and the tail, are black, as are also the quill feathers of the wings; and nearly all the rest of the body is white. The legs are green. The female is of a reddish brown, variously marked with black and dusky streaks. It is principally found in the western isles of Scotland, on the coasts of Norway, Iceland, and Greenland, and in many parts of North America.

The female lays from three to five eggs (sometimes as many as eight), which are large, smooth, glossy, and of a pale olive colour. She generally lays among stones, or plants, near the sea, but in a soft bed of down, which she plucks from her own breast. Sometimes two females will lay their eggs in the same nest, in which case they always agree remarkably well. As long as the female is sitting, the male continues on watch near the shore; but as soon as the young are hatched, he leaves them. The mother, however, remains with them a considerable time afterwards. It is curious to observe her manner of leading them out of the nest, almost as soon as they creep from the eggs. Going before them to the shore, they trip after her; and, when she comes to the waterside, she takes them on her back, and swims a few yards with them, when she dives; and the young ones are left floating on the surface, obliged to take care of themselves. They are seldom seen afterwards on land.

In Iceland, the Eider Ducks generally build their nests on small islands, not far from the shore; and sometimes even near the dwellings of the natives, who treat them with so much attention and kindness as to render them nearly tame. From these birds is produced the soft down, so well known by the name of eider, or edder down, which is so light and expansive that a couple of handfuls will fill a down quilt, which, in cold countries, is used instead of a quilt or blanket. In the breeding season the birds pluck it from their breasts to line their nests, and make a soft bed for the young ones. When the natives come to the nests, they carefully remove the female and take away the superfluous down and eggs; after this, they replace the female: she then begins to lay afresh, and covers her eggs with new down, which she also plucks from her body; when this is scarce, or she has no more left, the male comes to her assistance, and covers the eggs with his down, which is white, and easily distinguished from that of the female. When the young ones leave the nest, which is about an hour after they are hatched, it is once more plundered. The most eggs and the best down are got during the first three weeks of their laying; and it has generally been observed, that they lay the greatest number of eggs in rainy weather. One female, during the time of laying, generally gives half a pound of down; which, however, is reduced one half after it is cleaned. When pure it is sold in Lapland for two rix dollars a pound. The Iceland Company at Copenhagen generally export from Iceland about one thousand five hundred or two thousand pounds weight of this down, besides what is privately purchased by foreigners.

The Greenlanders kill these birds with darts; pursuing them in their little boats, watching their course by the air bubbles when they dive, and always striking them when they rise to the surface wearied. The flesh is valued as food, and their skins are made into warm and comfortable under garments.

THE WIDGEON.

THIS bird weighs about twenty-two ounces; it has a black nail at the end of the upper mandible of the bill, the other part of which is of a lead colour; the structure of the head and mouth very much resembles the common wild duck, only the head does not seem to be quite so large, in proportion to the body, which also appears of a finer shape, and the wings longer. The crown of the head towards the base of the bill is of a pale pink colour, inclining to a reddish white; the other parts of the head and neck are red; the sides of the body and the upper part of the breast are tinctured with a very fair, glossy, and beautiful claret colour, with a few small transverse lines of black. The feathers on the back are brown, the edges more pale or ash coloured; the scapular feathers, and those under the fore part of the wings, are finely variegated with small transverse black and white lines, beautifully dispersed like waves; the quill feathers are some of them brown, with white tips, others have their outward webs of a blackish purple; other parts, especially those beyond the covert feathers, of a lovely fine blue; some of the exterior feathers have their outward webs inclining to black, with a fine purple gloss upon the borders, on which there are a number of small light coloured spots; the rest of the wing feathers are of a beautiful party-coloured brown and white. The upper part of the tail is ash coloured; the under part, behind the vent, black. The legs and feet are of a dark lead colour, and the claws black. The young of both sexes are gray, and this hue they retain till February, when the plumage of the male begins to assume its variegated tints. He is said to retain his bright colours till the end of July, and then to become dark and gray, so as scarcely to be distinguished from the female.

Widgeons are common in Cambridgeshire, England, the Isle of Ely, &c. where the male is called the Widgeon, and the female the Whewer. They feed upon wild periwinkles, grass, weeds, &c. which grow at the bottom

of rivers and lakes.   Their flesh has a very delicious taste,
not inferior to teal, or wild ducks.

THE TEAL.

Thɪs is the smallest bird of the duck kind: it is common
in England in the winter months; and some imagine that
it breeds there as well as it does in France.   It does not
usually weigh more than twelve ounces; and it measures
about sixteen inches from the point of the bill to the tip
of the tail, and from the extremity of each wing, when extended, nearly two feet.   The bill is of a dark brown colour, the head is considerably lighter, inclining to a bay,
with a large white stripe over each eye, bending downwards, towards the back part of the head: the back and
sides under the wings are curiously varied with lines of
white and black.   The breast is of a dirty coloured yellow, interspersed with dusky transverse lines; the belly
more bright, with yellowish brown spots: the quill feathers of the wings are of a dusky brown, with white edges;
the covert feathers appear of a fine shining green, with their
tips white; the scapular feathers are more inclining to an
ash colour; the legs and feet are brown, the claws black.
These birds feed on cresses, chervil, and other weeds, and
also on seeds and some kinds of water insects.   The flesh

is a great delicacy, and has a less fishy taste than any other of the wild Duck tribe.

The female constructs her nests of reeds interwoven with grass, and is said to make it among rushes, that it may rise or fall with the varying height of the water.

These are the most common birds of the Duck kind in England; but who can describe the amazing variety of this tribe, if he extends his view to the different quarters of the world? The most noted of the foreign tribe are, the MUSCOVY DUCK, or, more properly speaking, the MUSK DUCK, so called from a supposed musky smell, with naked skin round the eyes, and which is a native of Africa. The BRAZILIAN DUCK, which is of the size of a goose, all over black, except the tips of the wings. The AMERICAN WOOD DUCK, with a variety of beautiful colours, and a plume of feathers, which falls from the back of the head like a friar's cowl.

The CHINESE or MANDARIN DUCK, is somewhat less than a widgeon, but remarkable for its elegance and beauty. The prevailing colour of its plumage on the upper parts is dusky brown; the scapulars, however, are black, and at the bend of the wing are three transverse streaks of black, and two of white alternately. The neck and breast are chestnut; the beak and legs are red, and the head is adorned with a fine expanded crest, the base of which is white, and the upper part of a beautiful glossy green.

These, and many others, might be added, were increasing the number of names the way to enlarge the sphere of our comprehension.

All these live in the manner of domestic Ducks, keeping together in flocks in the winter, and flying in pairs in summer; bringing up their young by the water side, and leading them to their food as soon as out of the shell.

Their nests are usually built among heath or rushes, not far from the water; and they lay twelve, fourteen, or more eggs before they sit: yet this is not always their method; the dangers they continually encounter from their situation sometimes oblige them to change their manner of building; and their awkward nests are often seen exalted on the tops of trees. This must be a very great labour to perform, as the Duck's bill is but ill formed for building a nest, and giving the materials of which it is composed a sufficient stability to stand the weather. The nest, whether high or low, is generally composed of the longest grass, mixed with heath, and is lined within with the bird's own feathers.

As these animals possess the faculties of flying and swimming, so they are in general birds of passage, and it is most probable perform their journeys across the ocean as well on the water as in the air. Those that migrate to England, on the approach of winter, are seldom found so well tasted or so fat as the fowls that continue there the year round: their flesh is often lean, and still oftener fishy; which flavour it has probably contracted in the journey, as their food in the lakes of Lapland, whence they descend, is generally of the insect kind.

As soon as they arrive in England, they are generally seen flying in flocks, to make a survey of those lakes where they intend to take up their residence for the winter. Lakes, with a marsh on one side, and a wood on the other, are seldom without vast quantities of wild fowl. The greatest quantities are taken in decoys; which, though well known near London, are yet untried in the remoter parts of the country. The manner of making and managing a decoy is as follows:—

A place is to be chosen for this purpose far remote from the common highway, and all noise of people. When the place is chosen, the pool, if possible, is to be planted round with willows, unless a wood answers the purpose of shading it on every side. On the south and north side of this pool are two, three, or four ditches or channels, made broad towards the pool, and growing narrower till they end in a point. These channels are to be covered over with nets supported by hooped sticks bending from one side to the other; so that they form a vault or arch growing narrower and narrower to the point, where it is terminated by a tunnel-net, like that in which fish are caught in wears. Along the banks of these channels, so netted over, which are called pipes, many hedges are made of reeds slanting to the edge of the channel, the acute angles to the side next the pool. The whole apparatus also is to be hidden from the pool by a hedge of reeds along the margin, behind which the fowler manages his operations. The place being fitted in this manner, the fowler is to provide himself with a number of wild Ducks made tame, which are called decoys. These are always to be fed at the mouth or entrance of the pipe, and to be accustomed to come at a whistle.

As soon as the evening is set in, the *decoy rises*, as they term it, and the wild fowl feed during the night. If the evening be still, the noise of their wings, during their flight, is heard at a very great distance, and produces no unpleasant sensation. The fowler, when he finds a fit opportunity, and sees his decoy covered with fowl, walks about the pool, and observes into what pipe the birds gathered in the pool may be enticed or driven. Then casting hemp seed, or some such seed as will float on the surface of the water, at the entrance and up along the pipe, he whistles to his decoy Ducks, who instantly obey the summons, and come to the entrance of the pipe, in hopes of being fed as usual. Thither also they are followed by a whole flock of wild ones, who little suspect the danger preparing against them. The wild Ducks, therefore, pursuing the decoy Ducks, are led into the broad mouth of the channel or pipe, nor have the least suspicion of the man, who keeps hidden behind one of the hedges. When they have got up the pipe, however, finding it grow more and more narrow, they begin to suspect danger, and would return back; but they are now prevented by the man, who shows himself at the broad end below. Thither, therefore, they dare not return; and rise they may not, as they are kept by the net above from ascending. The only way left them, therefore, is the narrow-funneled net at the bottom; into this they fly, and there they are taken.

It often happens, however, that the wild fowl are in such a state of sleepiness or dozing, that they will not follow the decoy ducks. Use is then generally made of a dog who is taught his lesson. He passes backward and forward between the reedhedges, in which there are little holes, both for the decoy man to see, and for the little dog to pass through. This attracts the eye of the wild fowl; who, prompted by curiosity, advance towards this little animal, while he all the time keeps playing among the reeds, nearer and nearer the funnel, till they follow him too far to recede. Sometimes the dog will not attract their attention till a red handkerchief, or something very singular, be put about him. The decoy Ducks never enter the funnel-net with the rest, being taught to dive under water as soon as the rest are driven in.

To this manner of taking wild fowl in England, we will subjoin another still more extraordinary, frequently practised in China. Whenever the fowler sees a number of Ducks settled in any particular plash of water, he sends off two or three gourds to float among them. These gourds resemble our pompions; but, being made hollow, they swim on the surface of the water; and on one pool there may sometimes be seen twenty or thirty of these gourds floating together. The fowl at first are a little shy at coming near them; but by degrees they come nearer; and as all birds at last grow familiar with a scarecrow, the Ducks gather about these, and amuse themselves by whetting their bills against them. When the birds are as familiar with the gourds as the fowler could wish, he then prepares to deceive them in good earnest. He hollows out one of these gourds large enough to put his head in; and making holes to breathe and see through, he claps it on his head. Thus accoutred, he wades slowly into the water, keeping his body under, and nothing but his head in the gourd above the surface; and in that manner moves imperceptibly towards the fowls, who suspect no danger. At last, however, he fairly gets in among them; while they, having been long used to see gourds, take not the least fright while the enemy is in the very midst of them; and an insidious enemy he is; for ever as he approaches a fowl, he seizes it by the legs, and draws it in a jerk under water. There he fastens it under his girdle, and goes to the next, till he has thus loaded himself with as many as he can carry away. When he has got this quantity, without ever attempting to disturb the rest of the fowls on the pool, he slowly moves off again; and, in this manner, pays the flock three or four visits in a day. Of all the various artifices for catching fowl, this seems likely to be attended with the greatest success, and is the most practised in China.

AMERICAN DUCKS.

WE have extracted the following list of Ducks, found in America, from Wilson's Ornithology.

The EIDER DUCK is found on the American as well as

the European shores, from 45° north to the highest latitudes yet discovered. It is said to breed as far south as Portland in Maine.

THE BLACK, OR SURF DUCK is a fine large species peculiar to America, and confined to the shores and bays of the sea.

THE VELVET DUCK is sometimes confounded with the Black Duck, but is smaller. It is common on the northern shores of Europe, and on those of Kamtschatka.

THE SCOTU DUCK abounds on the northern shores of both continents.

THE RUDDY DUCK appears to have recently made its appearance on the shores of the Middle States, and resembles the Ural Duck of Europe. Its colour is that of bright mahogany.

THE CANVASS-BACK DUCK is a very celebrated species unknown in Europe. They appear in the United States about the middle of October, and great numbers of them are found on the rivers near Chesapeake Bay. The Canvass-back, in the rich juicy tenderness of its flesh, and its delicacy and flavour, stands unrivalled by the whole of its tribe in perhaps any other part of the world. They sometimes sell from one to three dollars a pair. Its length is about two feet, and its weight two pounds.

THE RED-HEADED DUCK is a common associate of the Canvass-back, and its flesh is very little inferior. It is perhaps the Red-headed Widgeon of Europe.

THE SCAUP DUCK is common to both continents, and feeds on shell fish called Scaup.

THE TUFTED DUCK is a short plump bird, supposed by Wilson to be the same as the European bird of that name; but Charles Bonaparte has shown it to be a distinct species.

THE GOLDEN EYE is well known in Europe and in various parts of the United States. It may be easily known by the whistling of its wings as it passes through the air.

THE BUFFEL-HEADED DUCK, called also the Butterbox or Butter-ball, is common on the sea shores, rivers and lakes of the United States.

THE LONG-TAILED DUCK is common to both continents. They are often called *Old Wives* in the United States; and on the Chesapeake they go by the name of South Southerly, from their cry.

THE PIED DUCK is a scarce species, and is found only in salt water. It appears to be peculiar to America.

THE HARLEQUIN DUCK is found in Europe as well as America. At Hudson's Bay, it is called the Painted Duck; in New England, the Lord.

THE PELICAN.

The GREAT WHITE PELICAN OF AFRICA is much larger than a swan. Its four toes are all webbed together; and its neck, in some measure, resembles that of a swan: but that singularity in which it differs from all other birds, is in the bill, and the great pouch underneath, which are wonderful, and demand a distinct description. This enormous bill is fifteen inches from the point to the opening of the mouth, which is a good way back behind the eyes. The base of the bill is somewhat greenish; but it varies towards the end, being of a reddish blue. To the lower edges of the under chap hangs a bag, reaching the whole length of the bill to the neck, which is said to be capable of containing fifteen quarts of water. This bag the bird has a power of wrinkling up into the hollow of the under chap; but, by opening the bill, and putting one's hand down into the bag, it may be distended at pleasure. It is not

covered |with feathers, but a short downy substance, as
smooth and soft as satin.   Tertre affirms, that this pouch
will hold as many fish as will serve sixty hungry men for
a meal.   Such is the formation of this extraordinary bird,
which is a native of Africa and America.   It was once also
known in Europe, particularly in Russia ; but it seems to
have deserted those coasts.

The plumage of the Pelican which is now in the Tower
of London, is, all but the quill feathers of the wings, which
are black, of an extremely light and delicate flesh colour,
varied only by occasional darker tinges.   Except on the
temples, which are naked and flesh coloured, the head and
upper part of the neck are clothed with a short down.

The upper mandible is of a dull yellow in the middle, with a reddish tinge towards the edges, and a blood red spot at its curved extremity ; and the pouch is of a bright straw colour.

In the island of Manilla the Pelicans are of a rose colour, and in America they are brown. They are all torpid and inactive to the last degree, so that nothing can exceed their indolence but their gluttony. It is only from the stimulations of hunger that they are excited to labour ; for otherwise they would continue always in fixed repose. When they have raised themselves about thirty or forty feet above the surface of the sea, they turn their head, with one eye downwards, and continue to fly in that posture. As soon as they perceive a fish sufficiently near the surface, they dart down upon it with the swiftness of an arrow, seize it with unerring certainty, and store it up in their pouch. They then rise again, though not without great labour, and continue hovering and fishing, with their head on one side, as before.

This work they continue, with great effort and industry, till their bag is full : and then they fly to land, to devour and digest, at leisure, the fruits of their industry. This, however, it would appear, they are not long performing ; for, towards night, they have another hungry call ; and they again, reluctantly, go to labour.

Sometimes, they are said to assemble in large numbers, to act in concert, and to manœuvre with great skill, for the purpose of securing an abundant prey. This they accomplish by forming a circular line, and gradually narrowing the included space, till the fishes are driven within a narrow compass. They then all plunge into the water at once, on a given signal, fill their pouches with the spoil, and then return to the land, to enjoy themselves at leisure.

Their life is spent between sleeping and eating. The

c2

female makes no preparation for her nest, nor seems to choose any place in preference to lay in, but drops her eggs on the bare ground, to the number of five or six, and there continues to hatch them.   Her little progeny, however, seem to call forth some maternal affections: for its young have been taken and tied by the leg to a post, and the parent bird has been observed for several days to come and feed them; remaining with them the greatest part of the day, and spending the night on the branch of a tree that hung over them.   By these means they became so familiar that they suffered themselves to be handled; and they very readily accepted whatever fish was given to them.   These they always put first into their pouch, and then swallowed them at leisure.

With all the seeming indolence of this bird, it is not entirely incapable of instruction in a domestic state.   Father Raymond assures us, that he has seen one so tame and well educated among the native Americans, that it would go off in the morning, at the word of command, and return before night to its master, with its great paunch distended with plunder; a part of which the savages would make it disgorge, and a part they would permit it to reserve for itself.

"The Pelican," as Faber relates, "is not destitute of other qualifications.   One which was brought alive to the Duke of Bavaria's court, where it lived forty years, seemed to be possessed of very uncommon sensations.   It was much delighted in the company and conversation of men, and in music, both vocal and instrumental; for it would willingly stand," says he, "by those that sung or sounded the trumpet; and stretching out its head, and turning its ear to the music, listened very attentively to its harmony, though its own voice was little pleasanter than the braying of an ass."

Gesner tells us, that the Emperor Maximilian had a tame Pelican, which lived for above eighty years, and which always attended his army on their march.

It was once believed that the Pelican feeds her young with her own blood; a fable for which we are indebted to some of the early fathers of the church, and which has been perpetuated by poets and heralds. The fact is, that the parent bird feeds the young by pressing its full pouch against its breast, and thus expelling a portion of the contents. The appearance of the bird when in this attitude, with the bloody spot on the end of its bill closely pressed against the delicate plumage of its breast, may, it has been well observed, readily account for the prevalence of such an idea in the minds of superficial observers.

THE FRIGATE PELICAN, OR MAN-OF-WAR BIRD

Is chiefly met with between the tropics. It is the size of a large fowl. The bill. is slender, five inches long, from the base of which a dark reddish skin spreads on each side of the head, and a large bag hangs down the throat; the whole plumage is brownish black, the tail is long, and much forked. It is often found above a hundred, and sometimes two hundred leagues from land, and sometimes settles on the masts of ships. Its amazing length of wing, which is not less than fourteen feet, enables it to take immense flights; and, when it is not successful in fishing, it attacks the gulls and other water-fowl, and makes them disgorge the fish which they have taken.

ROUGH BILLED PELICAN.

To such of our readers as have visited the estuaries of the Florida Coast, the demure and awkward attitude of this bird is perfectly familiar. In that portion of our country, this species occurs in large flocks, and they are often to be seen along the shores of the Mississippi and Missouri, imparting a peculiar character to the otherwise solitary scene; their solemn and quiet demeanour being in strict unison with the stillness of the uninhabited plains which surround them. They build in societies, and are seldom found except in flocks. When they are disturbed they rise in much confusion, but soon form in regular order, usually flying in long lines, though sometimes in a triangle, like geese, with their long bills resting on their breasts. Charles Bonaparte has confounded this bird with the brown species, from which however it appears to be distinct, both in appearance and habits.—*Cabinet of Nat. History.*

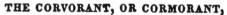

THE CORVORANT, OR CORMORANT,

Is about the size of a large Muscovy duck, and may be distinguished by its four toes being united by membranes together; and by the middle toe being toothed or notched, like a saw, to assist it in holding its fishy prey. The head and neck of this bird are of a sooty blackness, and the body thick and heavy, more inclining in figure to that of the goose than the gull. As soon as the winter approaches, they are seen dispersed along the sea shore, and ascending up the mouths of fresh water rivers, carrying destruction to all the finny tribe. They are most remarkably voracious, and have a most sudden digestion. Their appetite is forever craving, and never satisfied. This gnawing. sensation may probably be increased by the great quantity of small worms that fill their intestines, and which their increasing gluttony contributes to engender.

This bird has the most rank and disagreeable smell, and is more fœtid than even carrion, when in its most healthful state. It is seen as well by land as sea; it fishes in fresh water lakes, as well as in the depths of the ocean; it builds in the cliffs of rocks, as well as on trees; and preys not only in the day time, but by night.

Its indefatigable nature, and its great power in catching fish, were, probably, the motives that induced some nations to breed this bird up tame, for the purpose of fishing. The description of their manner of fishing is thus delivered by Faber.

"When they carry them out of the rooms where they are kept, to the fish pools, they hoodwink them, that they may not be frightened by the way. When they are come to the rivers, they take off their hoods; and having tied a leather thong round the lower part of their necks, that they may not swallow down the fish they catch, they throw them into the river. They presently dive under water; and there, for a long time, with wonderful swiftness, pursue the fish; and, when they have caught them, rise to the top of the water, and pressing the fish lightly with their bills, swallow them; till each bird has, after this manner, devoured five or six fishes. Then their keepers call them to the fist, to which they readily fly; and, one after another, vomit up all their fish, a little bruised with the first nip given in catching them. When they have done fishing, setting the birds on some high place, they loose the string

from their necks, leaving the passage to the stomach free
and open; and, for their reward, they throw them part of
their prey; to each one or two fishes, which they will
catch most dexterously, as they are falling in the air."

At present, the Corvorant is trained up in every part of
China for the same purpose. "It is very pleasant to behold with what sagacity they portion out the lake or the
canal where they are upon duty. When they have found
their prey, they seize it with their beak by the middle, and
carry it without fail to their master. When the fish is too
large, they then give each other mutual assistance: one
seizes it by the head, the other by the tail, and in this
manner carry it to the boat together. They have always,
while they fish, a string fastened round their throats, to
prevent them from devouring their prey." Such was formerly the practice in England; and as late as the reign of
Charles I. there was an officer of the household who bore
the title of Master of the Cormorants.

THE SHAG,

WHICH the French call the Lesser Corvorant, is another
of the pelican genus. The COMMON SHAG is in length
two feet. The general colour of its plumage is black, the
belly is dusky, and the head and neck glossed with green.
Like the corvorant it builds in trees. The CRESTED SHAG
is somewhat less than the preceding, and is less common.
The VIOLET and RED-FACED SHAGS are both natives of
Kamtschatka; and SPOTTED and CARUNCULATED SHAGS
are found in New Zealand. Besides these, there are
several other foreign species, particularly in Africa, where
there are two kinds of Shags not larger than a teal.

THE GANNET, OR SOLAN GOOSE,

Is of the size of a tame goose, but its wings much longer,
being six feet over. The bill is six inches long, straight

almost to the point. It differs from the corvorant in size,
being larger; in its colour, which is chiefly dirty white,
with a cinereous tinge; and by its having no nostrils, but

in their place a long furrow that reaches almost to the end
of the bill. From the corner of the mouth is a narrow slip
of black bare skin, that extends to the hind part of the
head; beneath the skin is another that, like the pouch of
the Pelican, is dilitable, and of size sufficient to contain
five or six entire herrings, which in the breeding season it
carries at once to its mate or its young.

These birds, which subsist entirely upon fish, chiefly
resort to those uninhabited islands where their food is
found in plenty, and men seldom come to disturb them.
The islands to the north of Scotland, the Skelig islands
off the coasts of Kerry, in Ireland, and those that lie in the
north sea off Norway, abound with them. But it is on the
Bass Island, in the Firth of Edinburgh, where they are
seen in the greatest abundance. " It is scarcely possible
to walk there without treading on them: the flocks of
birds upon the wing are so numerous as to darken the air
like a cloud; and their noise is such, that one cannot, without difficulty, be heard by the person next to him."

The Gannet is a bird of passage. In winter it seeks the more southern coasts of Cornwall, in England, hovering over the shoals of herrings and pilchards that then come down from the northern seas: its first appearance in the northern islands is in the beginning of spring; and it continues to breed till the end of summer. But, in general, its motions are determined by the migrations of the immense shoals of herrings that come pouring down at that season through the British Channel, and supply all Europe as well as this bird with their spoil. The Gannet assiduously attends the shoal in their passage, keeps with them in their whole circuit round the island, and shares with the fishermen this exhaustless banquet. As it is strong of wing, it never comes near the land, but is constant to its prey. The young Gannet is accounted a great dainty by the Scots, and sold very dear.

These birds breed but once a year, on the highest and steepest rocks near the sea, and lay only one egg, but if that be taken away, they lay another; and if that be also taken away, then a third; but never more for that season. Their eggs are white, and rather less than those of the common goose; and their nest large, composed of such substances as are found floating on the surface of the sea. The young birds during the first year differ greatly in colour from the old ones; being of a dusky hue, speckled with numerous triangular white spots.

These birds, when they pass from place to place, unite in small flocks of from five to fifteen; and, except in very fine weather, fly low, near the shore, but never pass over it; doubling the capes and projecting parts, and keeping nearly at an equal distance from the land. During their fishing they rise high into the air, and sail aloft over the shoals of herrings or pilchards, much in the manner of kites. When they observed the shoal crowded thick to-

gether, they close their wings to their sides and precipitate themselves head foremost into the water, dropping
almost like a stone. Their eye in this act is so correct,
that they never fail to rise with a fish in their mouth. If
in flying away with one, they see another they like better,
they immediately drop the first to seize it. The force
with which it descends on its prey may be imagined from
a circumstance which occurred, some years ago, at Pen-
zance, in Cornwall. As some pilchards were lying on a
fir plank, a Gannet darted down on them with such impetuosity as to strike its bill quite through the board, which was
an inch and a quarter thick, and, as may be supposed, to
kill itself on the spot. The St. Kildans sometimes take
them by fastening a herring to a plank, and setting it
afloat. The bird swoops down, and breaks its neck by
its violence.

THE BOOBY

Is also a species of the Pelican. The upper parts of the
plumage are brown ; the breast and belly white. It is found
in several parts of America, and is described as a very
simple bird.

THE ALBATROSS

Is one of the largest and most formidable birds of Africa
and South America. The largest, which is called the
WANDERING ALBATROSS, is rather larger than a swan
and its wings, when extended, ten feet from tip to tip
The bill, which is six inches long, is yellowish, and terminates in a crooked point. The top of the head is of a
bright brown; the back is of a dirty, deep spotted brown;
and the belly, and under the wings, is white. The toes,
which are webbed, are of a flesh colour.

This bird is an inhabitant of the tropical climates, and

also beyond them, as far as the Straits of Magellan, in the South Seas. It not only eats fish, but also such small waterfowl as it can take by surprise. It preys, as the gull kind do, upon the wing, and chiefly pursues the flying fish that are forced from the sea by the dolphins.

The Albatross seems to have a peculiar affection for the penguin, and a pleasure in its society. They are always seen to choose the same places of breeding; some distant, uninhabited island, where the ground slants to the sea, as the penguin is not formed either for flying or climbing. In such places their nests are seen together, as if they stood in need of mutual assistance and protection. In the middle, on high, the Albatross raises its nest on heath, sticks, and long grass, about two feet above the surface; and round this the penguins make their lower settlements, rather in holes in the ground; and most usually eight penguins to one Albatross.

There are about three other species of Albatross, all of them smaller than the preceding. The upper parts of the plumage are a dusky blue black, and the rump and under parts white; but what peculiarly distinguishes it is, that the bill, which is four inches long, is black, all but the upper ridge, which is yellow quite to the tip. It inhabits the South Seas within the tropics.

THE SKIMMER, OR CUTWATER,

Is twenty inches in length, and in breadth three feet seven inches. The bill is of a very singular structure, the upper chap or mandible being above an inch shorter than the under, and the upper shuts into it, as a razor into its handle. The base of the bill is red, the rest black; and on the sides are several furrows. The forehead, chin, and all the under parts, are white; the upper parts of the plumage black, with a bar of white across each wing.

The tail is short and forked. It inhabits all America; is commonly on the wing, and skims along the surface to catch the small fish on which it feeds. It is frequently known by the name of the Razor-bill.

THE PENGUIN GENUS

INCLUDES about nine species, which seem to hold the same place in the southern parts of the world as the auks do in the north, neither of them having ever been observed within the tropics. The wings of the larger species do not enable them to rise out of the water, but serve them rather as paddles to help them forward when they attempt to move swiftly, and in a manner walk along the surface of the water. Even the smaller kinds seldom fly by choice; they flutter their wings with the swiftest efforts, without making way; and though they have but a small weight of body to sustain, yet they seldom venture to quit the water, where they are provided with food and protection.

As the wings of the Penguin tribe are unfitted for flight, the legs are still more awkwardly adapted for walking. This whole tribe have all above the knee hid within the belly; and nothing appears but two short legs, or feet, as some would call them, that seem stuck under the rump, and upon which the animal is very awkwardly supported. They seem, when sitting, or attempting to walk, like a dog that has been taught to sit up or to move a minuet. Their short legs drive the body in progression from side to side; and were they not assisted by their wings, they could scarcely move faster than a tortoise.

This awkward position of the legs, which so unqualifies them for living upon land, adapts them admirably for a residence in water; in that, the legs placed behind the moving body, push it forward with greater velocity; and these birds, like Indian canoes, are the swiftest in the water, by having their paddles in the rear.

They are also covered more warmly all over the body with feathers than any other birds whatever; so that the sea seems entirely their element.

The PATAGONIAN PENGUIN weighs about forty pounds, and is four feet three inches in length. The bill measures four inches and a half, but is slender. The head, throat, and hind part of the neck, are brown; the back of a deep ash colour; and all the under parts white. The MAGELLANIC PENGUIN is about the size of a goose; the upper parts of the plumage are black, and the under white. These birds walk erect with their heads on high, their finlike wings hanging down like arms; so that to see them at a distance, they look like so many children with white aprons. Hence they are said to unite in themselves the qualities of men, fowls, and fishes. Like men, they are upright; like fowls, they are feathered; and, like fishes, they have fin-like instruments, that beat the water before, and serve for all the purposes of swimming rather than flying.

There are CRESTED PENGUINS at Falkland's Island, which are very beautiful birds. This is sometimes called the Hopping Penguin, or Jumping Jack, from the circumstance of its leaping quite out of the water, often to the height of three or four feet, when it meets an obstacle to its course.

At New Zealand there is a species of Penguin, which is not larger than a teal.

All the species feed upon fish; and seldom come ashore, except in the breeding season. Their flesh is rank and fishy; though the sailors say, that it is pretty good eating. In some the flesh is so tough, and the feathers so thick, that they stand the blow of a scimitar without injury.

The Penguin lays but one egg; and, in frequented shores, is found to burrow like a rabbit: sometimes three or four take possession of one hole, and hatch their young together. The egg of the Penguin is very large ,for the size of the bird, that of the smaller sorts being generally found bigger than that of a goose.

THE TROPIC BIRD

INCLUDES only three known species, which are all distinguished by a wedge-like tail, the two middle feathers extending a vast length beyond the others.

The COMMON TROPIC BIRD is about the size of a widgeon. The length to the tip of the two long feathers is nearly three feet. The bill is three inches long, and red; the head, neck, and under parts of the body, are quite white: the upper parts of the plumage white also, but marked with black lines. The two middle feathers of the tail measure twenty inches, and project fifteen inches beyond the rest. It takes its name from being chiefly found within the tropics. It frequently flies very high, but generally attends upon the flying fishes in their escape from their watery enemies; and they have now and then been found in calm weather supinely floating on the backs of the drowsy tortoises. Their flesh is not good, but is sometimes eaten by the hungry sailors.

On Palmerston Island there is a BLACK-BILLED TROPIC BIRD; and at Mauritius there is a Tropic Bird with a bill and a tail of a beautiful rose colour.

THE DARTER

Is distinguished by a peculiarly long and slender neck, and includes three species.

The WHITE-BELLIED DARTER is scarcely so large as a mallard, but its neck is so long that it measures not less than two feet ten inches. The bill is three inches long, straight, and pointed. The neck is covered with downy soft feathers, of a reddish gray: the upper parts of the plumage are dusky black, dashed with white; the under parts pure silvery white. It is a native of Brazil, and is extremely expert at catching fish.

The BLACK-BELLIED DARTER* is the size of the common duck. The head, neck, and breast are light brown; the back, scapulars, &c. marked with stripes of black and white; the quill feathers, belly, thighs, tail, are deep black. The four toes are united like those of the corvorant. In the island of Ceylon and Java it sits on the shrubs that hang over the water, and in a country where people are so apprehensive of serpents, it often terrifies the passengers by darting out its long and slender neck, which in their surprise they mistake for the attack of some fatal reptile.

CHAP. VII.

Of Fishes in general...Of cetaceous Fishes...The WHALE*.., The Fin Fish...The Narwal, or Sea Unicorn...The* SPERMACETI WHALE*...The* DOLPHIN*...Grampus, Porpesse, &c.*

THE number of fish to which we have given names, and with the figure of which at least we are a little acquainted, is, according to Linnæus, above four hundred. The majority of these are confined to the sea, and would expire in the fresh water, though there are a few which annually swim up the rivers, to deposit their spawn.

Wonderful as it may appear to see creatures existing in a medium so dense that men, beasts, and birds must inev-

* This bird is an inhabitant of the Carolinas, Georgia, the Floridas, and Louisiana; and is common in Cayenne and Brazil.—*Wilson.*

itably perish in it, yet experience proves that, besides those species which we are in the daily habit of seeing, the very depths of the immense ocean contain myriads of animated beings, to whose very form we are almost strangers, and of whose dispositions and manners we are still more ignorant. It is probable, indeed, that the fathomless recesses of the' deep contain many kinds of fish that are never seen by man. In their construction, modes of life, and general design, the watery tribes are perhaps still more astonishing than the inhabitants of either the land or air.

The structure of fish, and their adaptation to the element in which they are to live, are eminent proofs of divine wisdom. Most of them have the same external form, sharp at each end, and swelling in the middle, by which configuration they are enabled to traverse their native element with greater ease and swiftness. From their shape, men originally took the idea of those vessels which are intended to sail with the greatest speed; but the progress of the swiftest sailing ship, with.the advantage of a favourable wind, is far inferior to that of fish. Ten or twelve miles an hour is no small degree of rapidity in the sailing of a ship; yet any of the larger species of fish would soon overtake her, play round as if she did not move, and even advance considerably before her.

The fins of fish are denominated from their situations. The pectoral fins are placed at a little distance behind the opening of the gills, and are large and strong; and serve as well to balance the body as to assist the motion of the fish. The ventral fins are placed towards the lower part of the body, under the belly, and serve chiefly to raise or depress the fish in the water. The dorsal fins are situated on the ridge of the back, and are very large in flat fish: their use, like the pectoral ones, is to keep the body in equilibrio, as well as to contribute to its progressive motion.

The anal fins are placed between the vent and the tail, enabling the fish to keep an upright position.

The chief instruments of a fish's motion are the fins, which in some fish are more numerous than in others. The fish, in a state of repose, spreads all its fins, and seems to rest upon its pectoral and ventral fins near the bottom: if the fish folds up (for it has the power of folding), either of its pectoral fins, it inclines to the same side ; folding the right pectoral fin, its body inclines to the right side ; folding the left fin, it inclines to that side in turn. When the fish desires to have a retrograde motion, striking with the pectoral fins in a contrary direction effectually produces it. If the fish desires to turn, a blow from the tail sends it about ; but if the tail strikes both ways, then the motion is progressive. In pursuance of these observations, if the dorsal and ventral fins be cut off, the fish reels to the right and left, and endeavours to supply its loss by keeping the rest of its fins in constant employment. If the right pectoral fin be cut off, the fish leans to that side ; if the ventral fin on the same side be cut away, then it loses its equilibrium entirely. When the tail is cut off, the fish loses all motion, and gives itself up to where the water impels it.

The senses of fishes are remarkably imperfect; and, indeed, that of sight is almost the only one which, in general, they may be truly said to possess. But this is, in some degree compensated by their astonishing longevity, several species being known to live more than a hundred years. Their longevity is still exceeded by their singular fecundity ; for a single cod, for instance, produces at a birth two-thirds as many young ones as there are inhabitants in all Great Britain, above nine millions. The flounder produces at once above a million, and the mackerel five hundred thousand.

The spawn continues in its egg state in some fishes

longer than in others, and this generally in proportion to their size. The young of the salmon continues in egg from December to April; the carp, three weeks; and the little gold-fish, from China, is produced still quicker. The young spawn are the prey of all the inhabitants of the water, even of their own parents; and scarcely one in a thousand escapes the numerous perils of its youth.

Such is the general picture of these heedless and hungry creatures; but there are some in this class, living in the waters, that are possessed of finer organs and higher sensations; that have all the tenderness of birds or quadrupeds for their young; that nurse them with constant care, and protect them from every injury. Of this class are the *Cetaceous* order, or the fishes of the whale kind. There are others, though not capable of nursing their young, yet that bring them alive into the world, and defend them with courage and activity. These are the *Cartilaginous* kinds, or those which have gristles instead of bones. But the fierce unmindful tribe we have been describing, that leave their spawn without any protection, are called the *Spinous* or bony kinds, from their bones resembling the sharpness of thorns.

OF CETACEOUS FISHES.

This tribe is composed of the Whale, the Cachalot, the Dolphin, the Grampus, and the Porpesse. All these resemble quadrupeds in their internal structure, and in some of their appetites and affections. Like quadrupeds, they have lungs, a midriff, a stomach, intestines, liver, spleen, bladder, and parts of generation; their heart also resembles that of quadrupeds, with its partitions closed up as in them, and driving red and warm blood in circulation through the body; and to keep these parts warm, the whole kind are also covered between the skin and the muscles with a thick coat of fat or blubber.

As these animals breathe the air, it is obvious that they cannot bear to be any long time under water. They are constrained, therefore, every two or three minutes, to come up to the surface to take breath, as well as to spout out through their nostril (for they have but one), that water which they sucked in while gaping for their prey.

But it is in the circumstances in which they continue their kind, that these animals show an eminent superiority. Other fish deposit their spawn, and leave the success to accident; these never produce above one young, or two at the most; and this the female suckles entirely in the manner of quadrupeds, her breasts being placed, as in the human kind, above the navel. Their tails also are different from those of all other fish: they are placed so as to lie flat on the surface of the water; while the other kinds have them, as we every day see, upright or edgeways. This flat position of the tail enables them to force themselves suddenly to the surface of the water to breathe, which they are continually constrained to do.

THE WHALE.

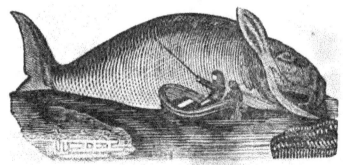

Of the Whale, properly so called, there are no less than seven different kinds; all distinguished from each other by their external figure or internal conformation. The GREAT GREENLAND WHALE, without a back fin, and black on the back; the ICELAND WHALE, without a back fin, and whitish on the back; the NEW ENGLAND WHALE, with a hump on the back; the WHALE WITH SIX HUMPS on the back; the FIN FISH, with a fin on the back near the tail; the PIKE-HEADED WHALE; and the ROUND-LIPPED WHALE. All these differ from each other in figure, as their names obviously imply. They differ also somewhat in their manner of living; the Fin Fish having a larger swallow than the rest; being more active, slender, and fierce; and living chiefly upon herrings.

The GREAT GREENLAND WHALE * is the fish, for taking which there are such preparations made in different parts of Europe. It is a large heavy animal, and the head alone makes a third of its bulk. It is usually found from sixty to seventy feet long. The fins on each side are from five to eight feet, composed of bones and muscles, and sufficiently strong to give speed and activity to the great mass of body which they move.

The tail is about twenty-four feet broad; and, when the fish lies on one side, its blow is tremendous. It is a cu-

---

* This whale occurs most abundantly in the frozen seas of Greenland and Davis' Straits, in the bays of Baffin and Hudson, in the sea to the northward of Behring's Strait, and along some parts of the northern shores of Asia and probably America. It is never met with in the German Ocean, and rarely within 300 leagues of the British coast; but along the coasts of Africa and South America, it is met with periodically in considerable numbers. In these regions it is attacked and captured by the Southern British and American whalers as well as by some of the people inhabiting the coasts, to the neighbourhood of which it resorts. Whether this Whale is precisely of the same kind as that of Spitzbergen and Greenland is uncertain, though it is evidently a mysticetus. One striking difference, possibly the effect of situation and climate, is, that the mysticetus, found in southern regions, is often covered with barnacles, while those of the Arctic seas are free from these shell fish.—*Godman.*

rious piece of mechanism, consisting of two lobes wholly made up of strong tendinous fibres, connected with the major part of the muscular structure of the body. Of those fibres there are three distinct layers, of which the two external are in the direction of the lobes, and the internal in an opposite direction. This structure renders the tail of the Whale one of the most flexible of animal organs. It can move all ways with equal ease; every part has its own individual motion.

The skin is smooth and black, and in some places marbled with white and yellow; which, running over the surface, has a very beautiful effect. The outward or scarf skin of the Whale is no thicker than parchment; but this removed, the real skin appears, of about an inch thick, and covering the fat or blubber that lies beneath: this is from eight to twelve inches in thickness; and is, when the fish is in health, of a beautiful yellow. The muscles lie beneath: and these, like the flesh of quadrupeds, are very red and tough.

The cleft of the mouth is above twenty feet long, which is near one third of the animal's whole length; and the upper jaw is furnisned with barbs, that lie, like the pipes of an organ, the greatest in the middle, and the smallest on the sides. These compose the whalebone, absurdly called fins, the longest spars of which are found to be not less than eighteen feet. The tongue is almost immoveably fixed to the lower jaw, seeming one great lump of fat; and, in fact, it fills several hogsheads with blubber. The eyes are not larger than those of an ox; and when the crystalline humour is dried, it does not appear larger than a pea. They are placed towards the back of the head, being the most convenient situation for enabling them to see both before and behind; as also to see over them, where their food is principally found. They are guarded by eyelids and eye-lashes, as in quadrupeds ; and they seem to be very sharp sighted.

Nor is their sense of hearing in less perfection; for they are warned, at great distances, of any danger preparing against them. We have already observed, that the substance, called whalebone, is taken from the upper jaw of the animal, and is very different from the real bones of the Whale. The real bones are hard, like those of great land animals, are very porous, and filled with marrow. Two great-strong bones sustain the under lip, lying against each other in the shape of a half-moon ; some of these are twenty feet long : they are often seen in gardens set up against each other, and are usually mistaken for the ribs.

The fidelity of these animals to each other exceeds whatever we are told of even the constancy of birds. Some fishers, as Anderson informs us, having struck one of two whales, a male and a female, that where in company together, the wounded fish made a long and terrible resistance : it struck down a boat with three men in it, with a single blow of the tail, by which all went to the bottom. The other still attended its companion, and lent it every assistance ; till, at last, the fish that was struck sunk under the number of its wounds; while its faithful associate, disdaining to survive the loss, with great bellowing, stretched itself upon the dead fish, and shared his fate.

The Whale goes with young nine or ten months, and is then fatter than usual, particularly when near the time of bringing forth. The young ones continue at the breast for a year ; during which time they are called by the sailors *short-heads*. They are then extremely fat, and yield above fifty barrels of blubber. The mother, at the same time, is equally lean and emaciated. At the age of two years they are called *stunts*, as they do not thrive much immediately after quitting the breast: they then yield scarce above twenty or twenty-four barrels of blubber: from that time forward they are called *skull-fish*, and their age is wholly unknown. The food of the Whale is a small insect which is seen floating in those seas, and which Linnæus terms the Medusa. These insects are black, and of the size of a small bean, and are sometimes seen floating in clusters on the surface of the water. They are of a round form, like snails in a box, but they have wings, which are so tender that it is scarce possible to touch them without breaking. These, however, serve rather for swimming than flying. They have the taste of raw muscles, and have the smell of burnt sugar. Inoffensive as the Whale is, it is not without enemies. There is a small animal, of the shell-fish kind, called the Whale-louse, that sticks to its body, as we see shells sticking to the foul bottom of a ship. This insinuates itself chiefly under the fins; and whatever efforts the great animal makes, it still keeps its hold, and lives upon the fat, which it is provided with instruments to arrive at.

The sword-fish, however, is the Whale's most terrible enemy. "At the sight of this little animal," says Anderson, "the Whale seems agitated in an extraordinary manner, leaping from the water as if with affright: wherever it appears, the Whale perceives it at a distance, and flies from it in the opposite direction. I have been myself," continues he, "a spectator of their terrible encounter. The Whale has no instrument of defence except the tail; with that it endeavours to strike the enemy; and a single blow taking place would effectually destroy its adversary: but the sword-fish is as active as the other is strong, and easily avoids the stroke; then bounding into the air, it falls upon its enemy, and endeavours not to pierce with its pointed beak, but to cut with its toothed edges. The sea all about is soon dyed with blood, proceeding from the wounds of the Whale; while the enormous animal vainly endeavours to reach its invader, and strikes with its tail against the surface of the water, making a report at each blow louder than the noise of a cannon."

There is still another and more powerful enemy, called, by the fishermen of New England, the killer. This is itself supposed to be a cetaceous animal, armed with strong and powerful teeth. A number of these are said to surround the Whale, in the same manner as dogs get round a bull. Some attack it with their teeth behind; others attempt it before; until, at last, the great animal is torn down, and its tongue is said to be the only part they devour when they have made it their prey. They are said to be of such great strength, that one of them alone was known to stop a dead Whale that several boats were towing along, and drag it from among them to the bottom.

But of all the enemies of these enormous fishes, man is the greatest: he alone destroys more in a year than the rest in an age, and actually has thinned their numbers in that part of the world where they are chiefly sought. At the first discovery of Greenland, Whales not being used to be disturbed, frequently came into the very bays, and were accordingly killed almost close to the shore; so that the blubber being cut off was immediately boiled into oil on the spot. The ships in those times took in nothing but the pure oil and the whalebone, and all the business was executed in the country; by which means a ship could bring home the product of many more Whales than she can according to the present method of conducting this trade. The fishery also was then so plentiful, that they were obliged sometimes to send other ships to fetch off the oil they had made, the quantity being more than the fishing ships could bring away. But time and change of circumstances have shifted the situation of this trade.   The ships coming in such numbers from Holland, Denmark, Hamburg, and other northern countries, all intruders upon the English, who were the first discoverers of Greenland, the Whales were disturbed, and gradually, as other fish often do, forsaking the place, were not to be killed so near the shore as before ; but are now found, and have been so ever since, in the openings and space among the ice, where they have deep water, and where they go sometimes a great many leagues from the shore.

The Whale fishery begins in May, and continues all June and July ; but whether the ships have good or bad success, they must come away, and get clear of the ice, by the end of August ; so that in the month of September at farthest they may be expected home : but a ship that meets with a fortunate and early fishery in May, may return in June or July.

The manner of taking Whales at present is as follows : —Every ship is provided with six boats, to each of which belongs six men for rowing the boat, and a harpooner, whose business it is to strike the Whale with his harpoon. Two of these boats are kept constantly on the watch at some distance from the ship, fastened to pieces of ice, and are relieved by others every four hours. As soon as a Whale is perceived, both the boats set out in pursuit of it, and if either of them can come up before the Whale finally descends, which is known by his throwing up his tail, the harpooner discharges his harpoon at him.   There is no difficulty in choosing the place where the Whale is to be struck, as some have asserted ; for these creatures only come up to the surface in order to spout up the water, or *blow*, as the fishermen term it, and therefore always keep the soft and vulnerable part of their bodies above water. As soon as the Whale is struck, the men set up one of their oars in the middle of the boat as a signal to those in the ship. On perceiving this, the watchman alarms all the rest with the cry of *fall! fall!* upon which all the other boats are immediately sent out to the assistance of the first.

The Whale finding himself wounded, runs off with prodigious violence. Sometimes he descends perpendicularly; at others goes off horizontally, at a small depth below the surface. The rope which is fastened to the harpoon is about two hundred fathoms long, and properly coiled up, that it may freely be given out as there is a demand for it. At first, the velocity with which this line runs over the side of the boat is so great, that it is wetted to prevent its taking fire: but in a short time the strength of the Whale begins to fail, and the fishermen, instead of letting out more rope, strive as much as possible to pull back what is given out already, though they always find themselves necessitated to yield at last to the efforts of the animal, to prevent his sinking their boat. If he runs out the two hundred fathoms of line contained in one boat, that belonging to another is immediately fastened to the end of the first, and so on ; and there have been instances, where all the rope belonging to the six boats has been necessary, though half that quantity is seldom required. The Whale cannot stay long below water, but again comes up to blow ; and being now much fatigued and wounded, stays longer above water than usual. This gives another boat time to come up with him, and he is again struck with a harpoon. He again descends, but with less force than before ; and when he comes up again, is generally incapable of descending, but suffers himself to be wounded and killed with long lances which the men are provided with for the purpose. He is known to be near death when he spouts up the water deeply tinged with blood.

The Whale, being.dead, is lashed alongside the ship.

They then lay it on one side, and put two ropes, one at the head, and the other in the place of the tail, which, together with the fins, is struck off as soon as he is taken, to keep these extremites above water. On the off side of the Whale are two boats, to receive the pieces of fat, utensils, and men, that might otherwise fall into the water on that side. These precautions being taken, three or four men with irons at their feet, to prevent slipping, get on the whale, and begin to cut out pieces of about three feet thick and eight long, which are hauled up at the capstan or windlass. When the fat is all got off, they cut off the whiskers of the upper jaw with an axe. Before they cut, they are all lashed to keep them firm; which also facilitates the cutting, and prevents them from falling into the sea: when on board, five or six of them are bundled together, and properly stowed; and after all is got off, the carcass is turned adrift, and devoured by the bears, who are very fond of it. In proportion as the large pieces of fat are cut off, the rest of the crew are employed in slicing them smaller, and picking out all the lean. When this is prepared, they stow it under the deck, where it lies till the fat of all the Whales is on board; then cutting it still smaller, they put it up in casks in the hold, cramming them very full and close. Nothing now remains but to sail homewards, where the fat is to be boiled and melted down into train oil.

A late improvement has been made in the method of discharging the harpoon, namely, by shooting it out of a kind of *swivel* or misquetoon; but it does not appear that since this improvement was made, the whale fishing ships have had better success than before.

The flesh of this animal is a dainty to some nations; and the savages of Greenland, as well as those near the south pole, are fond of it to distraction. They eat the flesh, and drink the oil, which is a first-rate delicacy. The finding a dead Whale is an adventure considered among

the fortunate circumstances of their lives. They make
their abode beside it; and seldom remove till they have
left nothing but the bones.

THE NARWAL, OR SEA-UNICORN,

SELDOM exceeds twenty-two feet long. Its body is slen-
derer than that of the whale, and its fat not in so great
abundance. But this great animal is sufficiently distinguished from all others of the deep, by its tooth or teeth,
which stand pointing directly forward from the upper jaw,
and are from nine to ten feet long. In all the variety of
weapons with which nature has armed her various tribes,
there is not one so large or so formidable as this. This
terrible weapon is generally found single; and some are
of opinion that the animal is furnished with but one by nature; but there is at present the skull of a Narwal at the
Stadthouse at Amsterdam, with two teeth. The tooth, or,
as some are pleased to call it, the horn of the Narwal, is

as straight as an arrow, about the thickness of the small of a man's leg, wreathed in the manner we sometimes see twisted bars of iron: it tapers to a sharp point; and is whiter, heavier, and harder than ivory. It is generally seen to spring from the left side of the head directly forward in a straight line with the body; and its root enters into the socket above a foot and a half. Notwithstanding its appointments for combat, this long and pointed tusk, amazing strength, and matchless celerity, the Narwal is one of the most harmless and peaceful inhabitants of the ocean. It is seen constantly and inoffensively sporting among the other great monsters of the deep, no way attempting to injure them, but pleased in their company. The Greenlanders call the Narwal the forerunner of the whale; for wherever it is seen, the whale is shortly after sure to follow. This may arise as well from the natural passion for society in these animals, as from both living upon the same food, which are the insects described in the preceding section. These powerful fishes make war upon no other living creature; and, though furnished with instruments to spread general destruction, are as innocent and as peaceful as a drove of oxen. The Narwal is much swifter than the whale, and would never be taken by the fishermen but for those very tusks, which at first appear to be its principal defence. These animals are always seen in herds of several at a time; and whenever they are attacked, they crowd together in such a manner, that they are mutually embarrassed by their tusks. By these they are often locked together, and are prevented from sinking to the bottom. It seldom happens, therefore, but the fishermen make sure of one or two of the hindmost, which very well reward their trouble.

THE CACHALOT, OR SPERMACETI WHALE

Has several teeth in the under jaw, but none in the upper. As there are no less than seven distinctions among whales, so also there are the same number of distinctions in the tribe we are describing. This tribe is not, of such enormous size as the whale, properly so called, not being above sixty feet long and sixteen feet high. In consequence of their being more slender, they are much more active than the common whale; they remain a longer time at the bottom, and afford a smaller quantity of oil. As in the common whale the head was seen to make a third part of its bulk, so in this species the head is so large as to make one half of the whole. Their throats are much wider than those of the common whale, as may be judged from the fact, that the remains of sharks more than twelve feet long have been found in their stomachs. The Cachalot is as destructive among the lesser fishes as the whale is harmless; and can at one gulp swallow a shoal of fishes down its enormous gullet. Linnæus tells us that this fish pursues and terrifies the dolphins and porpesses so much, as often to drive them on shore.

But, how formidable soever this fish may be to its fellows of the deep, it is by far the most valuable, and the most sought after by man, as it contains two very valuable drugs, spermaceti and ambergris : the whole oil of this fish is very easily convertible into spermaceti. This is performed by boiling it with a lye of potash, and hardening it in the manner of soap. Candles are now made of it, which are substituted for wax, and sold much cheaper.

As to the ambergris which is sometimes found in this whale, it was long considered as a substance found floating on the surface of the sea ; but time, that reveals the secrets of the mercenary, has discovered that it chiefly belongs to this animal. The name, which has been improperly given to the former substance, seems more justly to belong to this ; for the ambergris is found in the place where the seminal vessels are usually situated in other animals. It is found in a bag of three or four feet long, in round lumps, from one to twenty pounds weight, floating in a fluid rather thinner than oil, and of a yellowish colour. There are never seen more than four at a time in one of these bags ; and that which weighed twenty pounds, and which was the largest ever seen, was found single. These balls of ambergris are not found in all fishes of this kind, but chiefly in the oldest and strongest.

The blunt-headed Cachalot is fifty-four feet in length. Its greatest circumference is just beyond the eyes, and is thirty feet. The upper jaw is five feet longer than the lower, which is ten feet. The head is above one third the size of the fish. The end of the upper jaw is blunt, and near nine feet high, the spout hole placed near the end of it. The teeth are placed in the lower jaw, twenty-three on each side, all pointing outwards, and in the upper jaw, opposite, are a number of holes to receive them when the mouth is closed ; they are about eighteen inches long.

The Spermaceti Cachalot is found in greatest abundance in the Pacific Ocean, where large numbers of them are annually killed by the American and other whalers, for the sake of their oil and spermaceti.

The Spermaceti Cachalot is gregarious, and herds are frequently seen containing two hundred or more individuals.

The mode of attacking these animals is as follows:— Whenever a number of them are seen, four boats, each provided with two or three lines, two harpoons, four lances, and a crew of six men, proceed in pursuit, and, if possible, each boat strikes or " fastens to " a distinct animal, and each crew kill their own. When engaged in distant pursuit, the harpooner generally steers the boat, and in such cases the proper boat steerer occasionally strikes, but the harpooner mostly kills it. If one Cachalot of a herd is struck, it commonly takes the lead and is followed by the rest. The one which is struck, seldom descends far under water, but generally swims off with great rapidity, stopping after a short course, so that the boat can be drawn up to it by the line, or be rowed sufficiently near to lance it. In the agonies of death, the struggles of the animal are truly tremendous, and the surface of the ocean is lashed into foam by the motions of the fins and tail. Tall jets of blood are discharged from the blowholes, which show that the wounds have taken mortal effect, and seeing this, the boats are kept aloof, lest they should be dashed to pieces by the violent efforts of the victim.

When a herd is attacked in this way, ten or twelve of the number are killed; those which are only wounded are rarely captured. After the Cachalot is killed, the boats tow it to the side of the ship, and if the weather be fine, and other objects of chase in view, they are again sent to the attack.

About three tons of oil are commonly obtained from a large Cachalot; from one to two tons are procured from a small one. A cargo, produced from one hundred Cachalots may be from 150 to 200 tons of oil, besides the spermaceti, &c.

The SMALL-EYED, or BLACK-HEADED, SPERMACETI WHALE is one of the most formidable monsters of the deep. It has an enormous dark coloured head, armed with twenty-one projecting teeth on each side of the jaw. In a full grown specimen these teeth are nine inches in length. This Whale is often more than fifty feet in length, and is uncommonly active. Sharks, dolphins, and porpesses fall an easy prey to it.

THE GRAMPUS, THE DOLPHIN, AND THE PORPESSE.

ALL these fish have teeth both in the upper and lower jaw, and are much less than the whale. The Grampus, which is the largest, seldom exceeds twenty-four feet in length. It is a clumsy, unsightly fish, dark on the upper part, but very white below. The lower jaw is considerably wider than the upper. The back fin sometimes measures six feet. The Grampus is an exceedingly voracious animal, which does not always spare even its own kind. Packs of them are said to attack the Greenland whale, like bull dogs, and tear off his flesh in masses, It, however, displays the utmost solicitude and affection for its young. Little oil is afforded by the Grampus. It floats deep in the water, and would seldom be caught, did not its eagerness for prey prompt it to rush into shallow waters, where it is killed, but not till it has made a desperate and formidable resistance.

THE DOLPHIN.*

THOUGH so often incorrectly painted as being of the shape of the letter S. the Dolphin is almost straight, the back being very slightly incurvated, and the body slender; the nose is long, narrow, and pointed, with a broad transverse band, or projection of the skin on its upper part. From the shape of the nose, the animal has been called the Sea-goose. The mouth is very wide, and has twenty-one teeth in the upper, and nineteen in the lower jaw, somewhat above an inch long, conic at the upper end, sharp pointed, and bending a little in. They are placed at a small distance from each other; so that when the mouth is shut, the teeth of both jaws lock into each other; the spout hole is placed in the middle of the head; the tail is semi-lunar; the skin is smooth; the colour of the back and sides dusky; the belly whitish; it swims with great swiftness, and its prey is fish, but particularly cod, herrings, and flat fish. The Dolphin is longer and more slender than the porpesse, measuring nine or ten feet in length, and two in diameter.

All this species have fins on the back; very large heads; and resemble each other in their appetites, their manners, and conformation, being equally voracious, active, and roving. No fish could escape them, but from the awkward position of their mouth, which is placed in a manner under

* This is a distinct animal from the small fish which sailors call by the same name.

the head. Their own agility is so great as to prevent
them from being often taken, and they seldom remain a
moment above water; their too eager pursuit after prey
occasionally, however, exposes them to danger, as they
will sometimes follow the object of their pursuit even into
the nets of the fishermen.

A shoal of Dolphins will frequently attend the course of
a ship for the scraps that are thrown overboard, or the
barnacles adhering to their sides. A shoal of them followed the ships of Sir Richard Hawkins upwards of a
thousand leagues. Their gambols and evolutions on the
surface of the water are often very amusing.* A Dolphin

---

* Occasionally a troop of them may be seen scudding along, rising
in quick succession as if anxious each to get in advance of the other :
while again, a single individual may be observed successively rising
and falling in the same way, as if engaged in the act of catching a
prey.

In this way, shoals of Dolphins may be seen almost every day, and
at any hour feeding or sporting in the bay and rivers near the city of
New York, where we have sometimes enjoyed an opportunity of observing from the wharf, a large shoal of them moving down the Hudson
river with the tide : some plunging along as if in haste, others apparently at play, and others very slowly rising to the surface for breath, and
as gradually disappearing, allowing their dorsal fin to remain for a
considerable time above the surface.

The appearance of a shoal of these animals, at sea, moving in the
same direction, is considered by experienced mariners as an indica-
tion of an approaching storm, which very certainly follows their appearance. Falconer, in his beautiful poem of the Shipwreck thus
describes such a circumstance.

> "Now to the north from burning Afric's shore,
> A troop of porpesses their course explore ;
> In curling wreaths they gambol on the tide,
> Now bound aloft, now down the billow glide :
> Their tracks awhile the hoary waves retain
> That burn in sparkling trails along the main—

has been known to spring forward more than twenty feet
at a single bound. They inhabit the European and Pacific
ocean.

The flesh, though tolerably well tasted, is dry and insipid: the best parts are near the head. It is seldom eaten
but when young and tender. Dolphins are said to change
their colour before they die, and again after they are dead.

Many fabulous stories have been told of the Dolphin,
which has often been thus represented:

THE PORPESSE.

In its general form the Porpesse, or Porpus, very much resembles the Dolphin. It is, however, somewhat less in
size, and has a snout much broader and shorter. It is
generally from six to seven feet in length; its body is
thick towards the head, but grows slender towards the tail,
forming the figure of a cone. In each jaw are fortyeight teeth, small, sharp-pointed, and moveable; and so
placed that the teeth of one jaw lock into those of the other.
The eyes are small, as is the spout-hole at the top of the
head. In colours the back is black, and the belly whitish,

These fleetest coursers of the finny race,
When threatening clouds th' ethereal vault deface,
Their rout to leeward still sagacious form,
. To shun the fury of the approaching storm."—Canto II. § II.
Godman.

but they sometimes vary.—Porpesses are very numerous in all the British seas, but more particularly in the river St. Lawrence, in America; where there is a white kind. They are seldom seen except in troops of six or seven to thirty and upwards, and, like the dolphin, they are very agile and sportive. In the most tempestuous weather they can surmount the waves, and pursue their course without injury. Seamen have a superstitious detestation of them, because they believe their appearance to be ominous of approaching storms.

These animals live chiefly on the smaller fish: at the season when mackerel, herrings, pilchards, and salmon appear, the Porpesse swarms; and such is its violence in pursuit of its prey, that it will follow a shoal of small fish up a fresh water river, from whence it finds a difficulty to return. These creatures have been often taken in the river Thames, both above and below London Bridge; and it is curious to observe with what dexterity they avoid their pursuers, and how momentarily they recover their breath above the water. It is usual to spread four or five boats over the part of the river where they are seen, and to fire at them the instant they rise. One Porpesse yields about a hogshead of oil, and therefore renders its capture an object of consideration.

It is said that, whenever a Porpesse happens to be wounded, all the rest of its companions will immediately fall upon and devour it.

CHAP. VIII.

Of Cartilaginous Fishes...Of the SHARK...*The White Shark...
The Blue, the Long-tailed, the Basking, the Hammer-head-
ed, and the Angel Shark...The Remora...The Pilot Fish
...The Dogfish and its Varieties...The Sawfish...The* RAY...
*The Skate...The Rough-Ray...The Thornback...Manner
of fishing for Flat Fish...The Sting Ray...The Torpedo...
The Sea Devil...The Sea Eagle...The* LAMPREY...*The*
STURGEON...*The Isinglass Fish...The* ANGLER...*The* DI-
ODON, *or Sun Fish...The oblong and short Diodon...The*
SEA PORCUPINE...*The* LUMP-SUCKER...*The Sea Snail...
The* PIPE-FISH...*The* HIPPOCAMPUS...*The* GALLEY FISH.

CARTILAGINOUS FISHES.

THE first great distinction which the cartilaginous tribe of
fishes exhibits is, in having cartilages or gristles instead of
bones. The size of all fishes increases with age; but
from the pliancy of the bones in this tribe, they seem to
have no bounds placed to their dimensions: and it is supposed that they grow larger every day till they die.

Cartilaginous fishes unite the principal properties of both
the other classes in their conformation: like the cetaceous
tribes, they have organs of hearing, and lungs: like the
spinous kinds, they have gills, and a heart without a parti-
tion.

From this structure of their gills, these animals are enabled to live a longer time out of water than other fishes.
The cartilaginous shark, or ray, live some hours after they
are taken; while the spinous herring or mackerel expire
a few minutes after they are brought on shore. Some of
this class bring forth their young alive; and some bring
forth eggs, which are afterwards brought to maturity. In

all, however, the manner of gestation is nearly the same ;
for upon dissection, it is ever found, that the young, while
in the body, continue in the egg till a very little time before they are excluded: these eggs, they may properly be
said to hatch within their body ; and as soon as their young
quit the shell, they begin to quit the womb also.

THE SHARK.

Of all the inhabitants of the deep those of the Shark kind
are the fiercest and most voracious.

THE WHITE SHARK

Is sometimes seen to rank even among the whales for magnitude; and is found from twenty to thirty feet long.
Some assert that they have seen them of four thousand
pounds weight; and we are told particularly of one, that
had a human corpse in his belly. The head is large, and
somewhat flatted; the snout long, and the eyes large.
The mouth is enormously wide, as is the throat, and capable of swallowing a man with great ease. But its furni-
ture of teeth is still more terrible. Of these there are six
rows extremely hard, sharp pointed, and of a wedgelike fig-

ure. It is asserted that there are seventy-two in each jaw,
which make one hundred and forty-four in the whole;
yet others think that their number is uncertain; and that,
in proportion as the animal grows older, these terrible instruments of destruction are found to increase. With these
the jaws both above and below appear planted all over:
but the animal has the power of erecting or depressing
them at ,pleasure. When the Shark is at rest, they lie
quite flat in his mouth; but when he prepares to seize his
prey, he erects all this dreadful apparatus, by the help of
a set of muscles that join them to the jaw; and the animal
he seizes, dies, pierced with a hundred wounds, in a moment.

Nor is this fish less terrible to behold as to the rest of
his form: his fins are larger, in proportion; he is furnished
with great goggle eyes, which he turns with ease on every side, so as to see his prey behind him as well as before;
and his whole aspect is marked with a character of malignity: his skin also is rough, hard, and prickly; being that
substance which covers instrument cases, called shagreen.

No fish can swim so fast as the Shark; he outstrips the
swiftest ships. Such amazing powers, with such great appetites for destruction, would quickly unpeople even the
ocean; but providentially the Shark's upper jaw projects so
far above the lower, that he is obliged to turn on one side
(not on his back, as is generally supposed) to seize his prey.
As this takes some small time to perform, the animal pursued seizes that opportunity to make his escape.

Still, however, the depredations he commits are frequent
and formidable. The Shark is the dread of sailors in all
hot climates; where, like a greedy robber, he attends the
ships, in expectation of what may drop overboard. A man
who unfortunately falls into the sea at such a time is sure
to perish. A sailor that was bathing in the Mediterranean, near Antibes, in the year 1744, while he was swimming

about fifty yards from the ship, perceived a monstrous fish making towards him and surveying him on every side, as fish are often seen to look round a bait. The poor man, struck with terror at its approach, cried out to his companions in the vessel to take him on board. They accordingly threw him a rope with the utmost expedition, and were drawing him up by the ship's side, when the Shark darted after him from the deep, and snapped off his leg.

Mr. Pennant tells us, that the master of a Guinea ship, finding a rage for suicide prevail among his slaves, from a notion the unhappy creatures had, that after death they should be restored again to their families, friends, and country; to convince them at least that some disgrace must attend them here, he ordered one of their dead bodies to be tied by the heels to a rope, and so let down into the sea; and though it was drawn up again with great swiftness, yet, in that short space, the Shark had bitten off all but the feet. A Guinea captain was, by

stress of weather, driven into the harbour of Belfast, with
a lading of very sickly slaves, who, in the manner above-
mentioned, took every opportunity to throw themselves
overboard when brought upon deck, as is usual, for the
benefit of the fresh air. The brutal captain perceiving,
among others, a woman slave attempting to drown herself, pitched upon her as an example to the rest. As he
supposed that they did not know the terrors attending
death, he ordered the woman to be tied with a rope under
the armpits, and to let her down into the water When the
poor creature was thus plunged in, and about half way
down, she was heard to give a terrible shriek, which at
first was ascribed to her fears of drowning ; but soon after,
the water appearing red all round her, she was drawn up,
and it was found that a Shark, which had followed the ship,
had bit her off from the middle.

The usual method by which sailors take the shark, is by
baiting a great hook with a piece of beef or pork, which is
thrown out into the sea, by a strong cord, strengthened
near the hook with an iron chain. Without this precaution, the Shark would quickly bite the cord in two, and
thus set himself free. It is no unpleasant amusement to
observe this voracious animal coming up to survey the bait,
particularly when not pressed by hunger. He approaches
it, examines it, swims round it, seems for a while to neg-
lect it, perhaps apprehensive of the cord and chain : he
quits it for a little ; but, his appetite pressing, he returns
again; appears preparing to devour it, but quits it once
more. When the sailors have sufficiently diverted themselves with his different evolutions, they then make a pre-
tence, by drawing the rope, as if intending to take the
bait away ; it is then that the glutton's hunger excites him ;
he darts at the bait, and swallows it, hook and all. Sometimes, however, he does not so entirely gorge the whole,

but that he once more gets free; yet even then, though wounded and bleeding with the hook, he will again pursue the bait until he is taken. When he finds the hook lodged in his maw, his utmost efforts are then excited, but in vain, to get free: he tries with his teeth to cut the chain; he pulls with all his force to break the line; he almost seems to turn his stomach inside out, to disgorge the hook: in this manner he continues his formidable though fruitless efforts; till, quite spent, he suffers his head to be drawn above water, and the sailors, confining his tail by a noose, in this manner draw him on shipboard, and despatch him. This is done by beating him on the head till he dies: yet even that is not effected without difficulty and danger; the enormous creature, terrible even in the agonies of death, still struggles with his destroyers; nor is there an animal in the world that is harder to be killed. Even when cut in pieces, the muscles still preserve their motion, and vibrate for some minutes after being separated from the body. Another method of taking him, is by striking a barbed instrument, called a fizgig, into his body, as he brushes along by the side of the ship. As soon as he is taken up, to prevent his flouncing, they cut off the tail with an axe, with the utmost expedition.

This is the manner in which Europeans destroy the Shark; but some of the negroes along the African coast take a bolder and more dangerous method to combat their terrible enemy. Armed with nothing more than a knife, the negro plunges into the water, where he sees the Shark watching for his prey, and boldly swims forward to meet him. Though the great animal does not come to provoke the combat, he does not avoid it, and suffers the man to approach him; but, just as he turns upon his side to seize the aggressor, the negro watches the opportunity, plunges his knife into the fish's belly, and pursues his blows with such success, that he lays the ravenous tyrant dead at the bottom : he soon however returns, fixes the fish's head in a noose, and drags him to shore, where he makes a noble feast for the adjacent villages.

Nor is man alone the only enemy this fish has to fear: the REMORA, or Sucking-fish, is probably a still greater,

and follows the Shark every where. This fish has the power of adhering to whatever it sticks against, in the same manner as a cupping-glass sticks to the human body. It is by such an apparatus that this animal sticks to the Shark, drains away its moisture, and produces a gradual decay.

There are only three known species of the Sucking-fish ; these are occasionally seen in the Mediterranean Sea and the Pacific Ocean. The common Sucking-fish, which inhabits most parts of the ocean, is usually about a foot in length; the head large, equal in bigness to the body, which grows smaller gradually to the tail.

The back is convex and black, and the belly white. It has six fins, two growing from behind the gills, two more under the throat, a long one on the back, and opposite to it, under the belly, another of the same form and size ; the tail is wedge-shaped.

What this fish has peculiar to itself is, that the crown of the head is flat, and of an oval form, with a ridge, or rising, running lengthways ; and crossways to this, sixteen ridges, with hollow furrows between, by which structure it can fix to any animal or other substance, as they are often found adhering to the sides of ships, and the bodies of Sharks and other large fish. This adhesive quality gave rise to an absurd belief among the ancients, that the possessor of it had the power of arresting the progress of a ship in its fastest sailing. The Indians of Cuba and Jamaica formerly used to turn this quality to account by keeping tame Remoras, with which they fished. The Remora was secured by a slender but strong line, to which was attached a buoy, and was thrown into the water, upon which it would immediately pursue and fasten upon any fish that it perceived.

Sucking-fish are often eaten, and much admired: in taste they are said to resemble fried artichoke.

The Shark, however, appears to have one friend. This is the PILOT FISH. It has a long and banded body, with

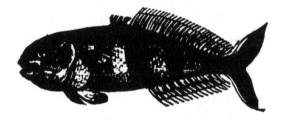

four loose spines on the back; a compresed head, rounded off in front; a small mouth, the jaws of which are of equal length, and furnished with small teeth. The palate has a curved row of teeth, and the tongue has teeth all along.

This species is found in the Mediterranean, Southern Ocean, East Indies, and Cape of Good Hope. It grows to a foot and a half in length, and derives its name from being commonly seen with the Shark, to which it appears to point out its prey. The circumstance of its guiding the Shark was long a matter of doubt, but appears now to be an ascertained fact. M. Geoffroy, when near Malta, in 1798, saw two of the Pilot Fish lead a Shark to a piece of bacon which a seaman had let down by a line and hook.

There are several other species of the Shark. The
BLUE SHARK is distinguished by a fine smooth skin on its
back, of a blue colour. The observation of Ælian, that
the young of this animal when pursued will take refuge
in the belly of its mother, by swimming down her mouth, is
confirmed by one of the best of modern icthyologists (Ron-
deletius). Mr. Pennant, however, does not apprehend
this circumstance to be peculiar to the Blue Shark, but rather common to the whole genus.

The LONG-TAILED SHARK. The author whom we have
just quoted mentions the dimensions of one of these animals which will serve to give an idea of the general proportions of this species. The fish in question was thirteen
feet in length, of which the tail was more than six, the
upper lobe much longer than the lower. The body
was round and short; the nose short and pointed; the
eyes large, and placed immediately over the corners of the
mouth. This fish was anciently called the *Sea Fox*, from
its supposed cunning.

The BASKING SHARK, which derives its name from its
basking on the surface of the water, has nothing of the rapacious nature of these animals, but feeds entirely on sea
plants, and some of the species of medusæ. They sometimes visit our coasts, in the summer season, when they
will lie in the sun on the surface of the water, and are so
tame as to suffer themselves to be stroked. They are in
length from three to twelve yards, and sometimes even
larger. The liver is of such immense size as often to
weigh nearly a thousand pounds. It contains a large
quantity of oil.

The HAMMER-HEADED SHARK OR BALANCE FISH, is
an animal of a very peculiar form. The head is placed
transversely to the body, like the head of a hammer or
mallet. It is terminated at each end by an eye, which is

so placed that it more conveniently looks downward than either upward or sideways. In the farther part of the forehead near the eyes, on each side there is a large oblong foramen or orifice, serving either for hearing or smelling, or perhaps for both. The mouth is very large, placed under the head, and armed with four rows of teeth extremely sharp. The tail consists of two fins, one longer than the other. The back is ash colour, and the belly white. This fish is chiefly caught in the Mediterranean.

The ANGEL SHARK, or MONK FISH, is the animal which connects the Shark genus with that of the ray, and partakes in some degree of the nature of both. It grows to a very large size, sometimes a hundred weight. The head is large; the teeth broad at the base, slender and sharp at the point. Like those of other Sharks, they are capable of being raised or depressed at the pleasure of the animal. The eyes are oblong, and placed lengthways in the head They are sunk very deep, and almost covered with the skin; and have more the expression of malevolence than of fire or spirit.

The skin is very rough; the back is of a pale ash colour, with a line of large lumps with pointed prickles along it. The pectoral fins are extremely large, and resemble wings, whence probably it derives the name of Angel. This species of Shark feeds on flounders and flat fish. It is extremely fierce and dangerous to be approached. Mr. Pennant speaks of a fisherman whose leg was terribly torn by one of them as it lay in his net in shallow water. It is not unfrequent on all our coasts.

In this genus are included the several species of DOG-FISH, which are common in most parts of the world, and retain much of the form and all the habits of the Shark. Nay, their appetite for human flesh is said to be so great, that they will sometimes even venture upon the shore to gratify this violent propensity. The most remarkable are the TOPE, which weighs sometimes seventy pounds; the SPOTTED DOG-FISH; the PICKED DOG-FISH, which has spines on its back fins; the SMOOTH HOUND, which is without the spines; and the PORBEAGLE.

Authors have classed under this genus a singular fish which is well known in the Western Ocean under the name of the SAW-FISH. It is remarkable for a curious instrument with which it is furnished at the snout resembling a saw, and which is sometimes of the length of five feet. From this circumstance it is evident that it must grow to a very large size.

THE RAY.

THE whole of this genus resemble each other very strongly in their figure; nor is it easy without experience to distinguish one from another. The stranger to this dangerous tribe may imagine he is only handling a skate when he is instantly struck numb by the torpedo; and he may suppose he has caught a thornback till he is stung by the fire-flare. It is by the spines that these animals are distinguished from each other.

THE SKATE

THIS fish is the largest and best of its tribe; the flesh being white, firm, and well flavoured. It sometimes attains an immense size. It has a broad flat body, brown on the back, and white on the belly. The principal difference between it and the thornback consists in its having sharp teeth, and a single row of spines on the tail, whilst the latter has blunt teeth, and several rows of spines both on the back and tail. The females produce their offspring from May till September. Each of the young ones is enclosed in an angular oblong bag, of a maroon colour; a substance like thin parchment or leather, and having two horns at each end. These, which are sometimes cast ashore after storms, are called purses by the fishermen.

The SHARP-NOSED RAY has ten spines, that are situated towards the middle of the back. The ROUGH RAY has its spines spread indiscriminately over the whole back. The THORNBACK has its spines disposed in three rows upon the back. The STING RAY, or FIRE-FLARE, has but one spine but indeed a terrible one. This dangerous weapon is placed on the tail, about four inches from the body, and is not less than five inches long. It is of a flinty hardness, the sides thin, sharp-pointed, and closely and sharply bearded the whole way. The TORPEDO has no spines that can wound; but in the place of them it is possessed of one of the most potent and extraordinary faculties in nature.

Of all the larger fish of the sea, these are the most numerous; and they owe their numbers to their size. Except the white shark and cachalot alone, there is no other fish that has a swallow large enough to take them in; and their spines make them a still more dangerous morsel. Yet the size of some is such, that even the shark himself is unable to devour them: we have seen some of them in England weigh above two hundred pounds; but that is nothing to their enormous bulk in other parts of the world. Labat tells us of a prodigious Ray that was speared by the negroes at Guadaloupe, which was thirteen feet eight inches broad, and about ten feet from the snout to the insertion of the tail. The tail itself was in proportion, for it was no less than fifteen feet long, twenty inches broad at its insertion, and tapering to a point. The body was two feet in depth; the skin as thick as leather, and marked with spots, which spots in all of this kind are only glands, that supply a mucus to lubricate and soften the skin. This enormous fish was utterly unfit to be eaten by the Europeans; but the negroes chose out some of the nicest bits, and carefully salted them up as a most favourite provision.

It is chiefly during the winter season that our fishermen fish for the Ray; but the Dutch, who are indefatigable, begin their operations earlier, and fish with better success than we do. The method practised by the fishermen of Scarborough is thought to be the best among the English; and, as Mr. Pennant has given a very succinct account of it, we shall present it to the reader.

"When they go out to fish, each person is provided with three lines: each man's lines are fairly coiled upon a flat oblong piece of wicker work; the hooks being baited and placed very regularly in the centre of the coil. Each line is furnished with two hundred and eighty hooks, at the distance of six feet two inches from each other. The hooks are fastened to lines of twisted horse-hair, twenty-seven inches in length. The line is laid across the current, and always remains upon the ground about six hours.

"The best bait for all kinds of fish is fresh herring cut in pieces of a proper size. Next to herrings are the lesser lampreys. The next baits in esteem are small haddocks cut in pieces, sandworms, muscles, and limpets; and, lastly, when none of these can be found, they use bullock's liver. The hooks used there are much smaller than those employed at Iceland and Newfoundland; and are two inches and

a half long in the shank. The line is made of small cord-
ing ; it is always tanned before it is used, and is in length
about three miles."

But this extent of line is nothing to what the Italians
throw out in the Mediterranean. Their fishing is carried,
on in a tartan, which is a vessel much larger than ours ;
and they bait a line of no less than twenty miles long,
with above ten or twelve thousand hooks. This line is
not regularly drawn every six hours, as with us, but remains for some time in the sea ; and it requires the space
of twenty-four hours to take it up. By this apparatus they
take Rays, sharks, and other fish ; some of which are above
a thousand pounds weight. When they have caught any
of this magnitude, they strike them through with a harpoon, to bring them on board, and kill them as fast as they
can.

This method of catching fish is obviously fatiguing and
dangerous ; but the value of the capture generally repays
the pains. The skate and the thornback are very good
food ; and their size, which is from ten pounds to two hundred weight, very well rewards the trouble of fishing for
them. But it sometimes happens that the lines are visited
by very unwelcome intruders ; by the Rough Ray, the Fire-
flare, or the Torpedo.

The Rough Ray inflicts but slight wounds with the
prickles with which its whole body is furnished. To the
ignorant it seems harmless, and a man would at first sight
venture to take it in his hand, without any apprehension ;
but he soon finds, that there is not a single part of its body
that is not armed with spines ; and that there is no way of
seizing the animal but by the little fin at the end of the
tail.

But this animal is harmless, when compared to the Sting
Ray, or Fire-flare, which seems to be the dread of even

the boldest and most experienced fishermen. . The spine, with which it wounds its adversaries, is not venomous, as has been vulgarly supposed, but is, in fact, a weapon of offence belonging to this animal, and capable, from its barbs, of inflicting a very terrible wound, attended with dangerous symptoms; it is fixed to the tail, as a quill is into the tail of a fowl, and is annually shed in the same manner.

THE TORPEDO

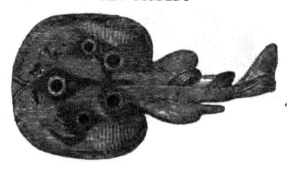

Is, however, the most remarkable of the Ray kind. The body of this fish is almost circular, and thicker than others of the same genus; the skin is soft, smooth, and of a dusky brown above, and white underneath; the eyes very small; the tail tapering to a point; and the weight of the fish from a quarter to fifteen pounds. Redi found one twenty-four pounds weight. The Electrical Rays are found in many parts of the European seas. The fishermen often discover it in Torbay, and sometimes of eighty pounds weight. They are partial to sandy bottoms, in about forty fathoms water, where they often bury themselves by flinging the sand over them, by a quick flapping of all the extremities. They bring forth their young in autumn. To all outward appearance, the Torpedo is furnished with no extraordinary qualities; yet such is the unaccountable power it possesses, that, the instant it is touched, it numbs not only the hand and arm, but sometimes also the whole body. The shock received resembles the stroke of an electrical machine ; sudden, tingling, and painful. It is, in truth, electric. "The instant," says Kempfer, "I touched it with my hand, I felt a terrible numbness in my arm, and as far up as the shoulder. Even if one treads upon it with the shoe on, it affects not only the leg, but the whole thigh upwards. Those who touch it with the foot, are seized with a stronger palpitation than even those who touch it with the hand. This numbness bears no resemblance to that which we feel when a nerve is a long time pressed and the foot is said to be asleep; it rather appears like a sudden vapour, which, passing through the pores, in an instant penetrates to the very springs of life, whence it diffuses itself over the whole body, and gives real pain. The nerves are so affected, that the person struck imagines all the bones of his body, and particularly those of the limb that received the blow, are driven out of joint. All this is accompanied with a universal tremor, a sickness of the stomach, a general convulsion, and a total suspension of the faculties of the mind."

Reaumur, who made several trials upon this animal, has convinced the world that it is not necessarily, but by an effort, that the Torpedo numbs the hand of him that touches it. He tried several times, and could easily tell when the fish intended the stroke, and when it was about to continue harmless. Always before the fish intended the stroke, it flattened the back, raised the head and the tail; and then, by a violent contraction in the opposite direction, struck with its back against the pressing finger; and the body, which before was flat became humped and round.

The electric or benumbing organs are placed one on each side of the gills, reaching from thence to the semicircular cartilages of each great fin, and extending longitudinally from the interior extremity of the animal to the transverse cartilage which divides the thorax from the abdomen, and within these limits they occupy the whole space between the skin of the upper and under surfaces. Each organ is about five inches in length, and at the anterior end, about three in breadth; they are composed of perpendicular columns, reaching from the upper to the under surface, varying in length according to the thickness of the parts of the body, from an inch and a half to half an inch. The engraving displays the interior of the lower electric or galvanic organ.

When the fish is dead, the whole power is destroyed, and it may be handled or eaten with perfect security. It is now known that there are more fish than this of the Ray kind possessed of the numbing quality, which has acquired them the name of the Torpedo.

There are two other species of Ray, which for their singular form deserve to be distinguished. The first is called the SEA DEVIL. Its nose and snout are divided, as it were, into two horns; and its sides are terminated by the fins. Its skin, towards the head, is variegated with dusky spots. It grows, sometimes, to the length of six or seven feet.

The SEA EAGLE is another species of this deformed tribe. It receives its name from its thin and expanded sides, which resemble the spread wings of an eagle. Its head, in some degree, resembles that of a toad: its eyes are large and prominent. It is generally found small, but is said sometimes to grow to a very large size.

THE LAMPREY.

THERE is a species of the Lamprey served up as a great delicacy among the modern Romans very different from ours. Whether theirs be the murena of the ancients, we shall not pretend to say; but there is nothing more certain than that our Lamprey is not.

The Lamprey known among us is differently estimated, according to the season in which it is caught, or the place where it has been fed. Those that leave the sea to deposit their spawn in fresh waters are the best; those that are entirely bred in our rivers, and that have never been at sea, are considered as much inferior to the former. Those that are taken in the months of March, April, or May, just upon their leaving the sea, are reckoned very good; those that are caught after they have cast their spawn, are found to be flabby, and of little value.

The Lamprey much resembles the eel in its general appearance, but is of a lighter colour, and rather a clumsier make. It differs, however, in the mouth, which is round, and placed rather obliquely below the end of the nose. It more resembles the mouth of a leech than an eel; and the animal has a hole on the top of the head, through which it spouts water, as in the cetaceous kind. There are seven holes on each side for respiration; and the fins are formed rather by a lengthening out of the skin, than any set of bones or spines for that purpose. As the mouth is formed resembling that of a leech, so it has a property resembling that animal, of sticking close to and sucking any body it is applied to. It is extraordinary the power they have of adhering to stones; which they do so firmly, as not to be drawn off without some difficulty. We are told of one that weighed but three pounds; and yet it stuck so firmly to a stone of twelve pounds, that it remained suspended at its mouth; from which it was separated with no small difficulty. As to the intestines of the Lamprey, it seems to have but one great bowel, running from the mouth to the vent, narrow at both ends, and wide in the middle.

So simple a conformation seems to imply an equal simplicity of appetite. In fact, the Lamprey's food is either slime and water, or such small water-insects as are scarcely perceivable. Perhaps its appetite may be more active at sea, of which it is properly a native; but when it comes up into our rivers, it is hardly perceived to devour any thing.

Its usual time of leaving the sea, which it is annually seen to do in order to spawn, is about the beginning of spring; and after a stay of a few months it returns again to the sea. Their preparation for spawning is peculiar; their manner is to make holes in the gravelly bottoms of rivers; and on this occasion their sucking power is particularly serviceable; for if they meet with a stone of a considerable size, they will remove it, and throw it out. Their young are produced from eggs, in the manner of flat fish; the female remains near the place where they are excluded, and continues with them till they come forth. She is sometimes seen with her whole family playing about her; and after some time she conducts them in triumph back to the ocean.

THE STURGEON

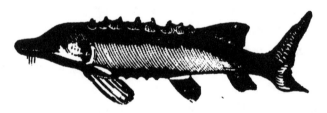

IN its general form resembles a fresh water pike. Formidable as this large and finely tasted fish is in its appearance, it is perfectly harmless; the body, which is from six to eighteen feet in length, is pentagonal, armed from head to tail with five rows of large bony tubercles, each of which ends in a strong recurved tip; one of these is on the back,

one on each side, and two on the margin of the belly.
The snout is long, and obtuse at the end, and has the tendrils near the tip. The mouth, which is beneath the head,
is somewhat like the opening of a purse, and is so formed
as to be pushed suddenly out, or retracted. The upper
part of the body is of a dirty olive colour; the lower parts
silvery; and the tubercles are white in the middle. The
tendrils on the snout, which are some inches in length,
have so great a resemblance in form to earth-worms, that,
at first sight, they might be mistaken for them. By this
contrivance, this clumsy, toothless fish is supposed to keep
himself in good condition, the solidity of his flesh evidently
showing him to be a fish of prey. He is said to hide his
body among the weeds near the sea-coast, or at the mouths
of large rivers, only exposing his tendrils, which small fishes or sea-insects, mistaking for real worms, approach to
seize, and are sucked into the jaws of their enemy. He
has been supposed by some to root into the soil at the
bottom of the sea or rivers; but, if this were the case, the
tendrils above mentioned, which hang from his snout over
his mouth, must be very inconvenient to him; as he has
no jaws, it is evident that he lives by suction, and, during
his residence in the sea, marine insects are generally
found in his stomach.

Of this fish there are three species, the COMMON STUR-
GEON, the CAVIAR STURGEON, and the HUSO, or ISINGLASS
FISH. The largest Sturgeon we have heard of caught in
Great Britain, was a fish taken in the Eske, where they
are most frequently found, which weighed four hundred
and sixty pounds. An enormous size to those who have
only seen our fresh water fishes!

As the Sturgeon is a harmless fish, and no way voracious, it is never caught by a bait in the ordinary manner
of fishing, but always in nets. From the quality of flound-

ering at the bottom it has received its name; which comes
from the German verb *stoeren*, signifying to wallow in
the mud.

The usual time for the Sturgeon to come up rivers to
deposit its spawn is about the beginning of summer, when
the fishermen of all the large European rivers make a regular preparation for its reception. At Pillau, in Russia, particularly, the shores are formed into districts, and allotted to
companies of fishermen, some of which are rented for
about three hundred pounds a year. The nets in which
the Sturgeon is caught, are made of small cord, and placed
across the mouth of the river; but in such a manner that,
whether the tide ebbs or flows, the pouch of the net goes
with the stream. The Sturgeon thus caught, while in the
water, is one of the strongest fishes that swims, and often
breaks the net to pieces that encloses it; but the instant
it is raised with its head above water, all its activity ceases;
it is then a lifeless, spiritless lump, and suffers itself to be
tamely dragged on shore.

The flesh of this animal pickled is very well known at
all the tables of Europe; and is even more prized in England than in any of the countries where it is usually caught.*
The fishermen have two different methods of preparing it.
The one is by cutting it in long pieces lengthwise, and having
salted them, by hanging them up in the sun to dry: the
fish thus prepared is sold in all the countries of the Levant,
and supplies the want of better provision. The other
method, which is usually practised in Holland, and along
the shores of the Baltic, is to cut the Sturgeon crosswise
into short pieces, and put it into small barrels, with a pickle

---

* The Sturgeon is found in the rivers of America, and is frequently
taken in the nets of the shad fishers. But it is not greatly esteemed in
this country, for the table.

made of salt and saumure. This is the Sturgeon which is
sold in England, and of which great quantities come from
the North.

A very great trade is also carried on with the roe of the
Sturgeon, preserved in a particular manner, and called caviar: it is made from the roe of all kinds of Sturgeon, but
particularly the second. This is much more in request in
other countries of Europe than in England. To all these
high relished meats, the appetite must be formed by degrees; and though formerly even in England it was very
much in request at the politest tables, it is at present sunk
entirely into disuse. It is still, however, a considerable
merchandise among the Turks, Greeks, and Venetians.
Caviar somewhat resembles soft soap in consistence; but
it is of a brown, uniform colour, and is eaten as cheese
with bread.

THE HUSO, OR ISINGLASS FISH,

FURNISHES a still more valuable commodity. This fish is
caught in great quantities in the Danube, from the months
of October to January: it is seldom under fifty pounds
weight, and often above four hundred; its flesh is soft, glutinous, and flabby; but it is sometimes salted, which makes
it better tasted, and then it turns red like salmon. It is
for the commodity it furnishes that it is chiefly taken. The
manner of making it is this; they take the skin, the entrails,
the fins, and the tail of this fish, and cut them into small
peices; these are left to macerate in a sufficient quantity
of warm water, and they are all boiled shortly after with a
slow fire, until they are dissolved and reduced to a jelly:
this jelly is spread upon instruments made for the purpose,
so that drying, it assumes the form of parchment, and, when
quite dry, it is then rolled into the form which we see it in
the shops. This valuable commodity is principally fur-

nished from Russia, where they prepare great quantities surprisingly cheap.

THE SPOTTED TOAD FISH, ANGLER, OR FISH-ING FROG,

In shape very much resembles a tadpole, or young frog, but of an enormous size, for it grows to above five feet long, and its mouth is sometimes a yard wide. The eyes are placed on the top of the head, and are encompassed with prickles; immediately above the nose are two long

beards or filaments, small in the beginning, but thicker at the end, and round; these, as it is said, answer a very singular purpose; for being made somewhat resembling a fishing-line, it is asserted, that the animal converts them to the purpose of fishing. With these extended, the Fishing-Frog is said to hide in muddy waters, and leave nothing but the beards to be seen; the curiosity of the smaller fish brings them to view these filaments, and their hunger induces them to seize the bait; upon which the animal in ambush instantly draws in its filaments with the little fish that had taken the bait, and devours it without mercy. This story, though apparently improbable, has found credit among some of our best naturalists. The fishermen have, in general, a great regard for this ugly fish, as it is an enemy to the dog fish, the bodies of those fierce and voracious animals being often found in its stomach: whenever they take it, therefore, they always set it at liberty. It is found in Brazil and China; and generally keeps at the bottom of the water, among sea-weed or between stones.

THE DIODON, OR SUN FISH,

Is easily distinguished by its very peculiar form; having a very deep body, and, as it were, cut off in the middle. There are three well known species.

The OBLONG DIODON grows to an immense size, and has been known to weigh upwards of five hundred pounds. In its form it resembles a bream or carp cut off in the middle. The mouth is very small, and contains in each jaw two broad teeth, with sharp edges. The dorsal and the anal fins are placed at the extremity of the body; the tail fin is narrow, and fills up all the space between these two fins. When boiled, it is observed to turn entirely to a glutinous jelly, and would probably serve all the purposes of isinglass; but it is not found in sufficient plenty at least upon our coasts.

The SHORT DIODON differs from the preceding, in being much shorter and deeper, resembling the head of a fish rather than a perfect animal; both kinds are found on the western coast of Britain, but in greater plenty in the warmer climates of Europe.

THE SEA PORCUPINE.

LIKE the porcupine, whence it takes its name, it is covered over with long thorns or prickles, which point on every side; and when the animal is enraged, it can blow up its body as round as a bladder, by means of a sort of air-bag in its interior. It varies in dimensions from the size of a football to that of a bushel. The back is of a bluish colour, the sides and belly are white, and the body is covered with light and dark brown spots. Of this extraordinary creature there are many species: some threatening only with spines, and others defended with a bony helmet that covers the head.

This species is found not only in America, but in the

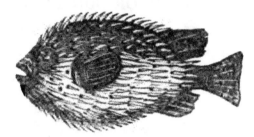

Red Sea, and on the Japanese shores. It is sometimes called the Goad Fish. The people catch them merely for amusement. They throw in a line baited with the tail of a sea crab; the fish approaches, but being afraid of the line, he makes several turns and trials round the bait, and at length nibbles at it, but pretends to reject it, and passes by, striking it with his tail, as if he did not regard it. But if the rod be kept steady, he presently turns back, seizes the bait, and swallows hook and all. When he finds himself taken, he becomes enraged, bristles up his spines, swells out his belly, and endeavours to wound every thing that is near him. Finding this of no avail, he resorts to cunning, and seems to submit: he lowers his spines, contracts his body, and lies like a wet glove. But this artifice not succeeding, and perceiving the fisherman dragging him towards the land, he renews his defensive attitude with redoubled fury. His spines are now vigorously erected, his form rounded, and his body so completely armed at all points, that it is impossible to take it by the hand; he is therefore dragged to some distance, where he struggles and quickly dies.

THE LUMP FISH, LUMPSUCKER, OR SEA OWL,

Is sixteen inches in length, and its weight about four pounds; the shape of the body is like that of the bream, deep, and it swims edgeways, the back is sharp and elevated, and the belly flat; the lips, mouth, and tongue of this animal are of a deep red; the whole skin is rough, with bony knobs, the largest row is along the ridge of the back; the belly is of a bright crimson colour; but what makes the chief singularity in this fish is an oval aperture in the belly, surrounded with a fleshy, soft substance, that seems bearded all round; by means of this part it adheres with vast force to any thing it pleases. If flung into a pail of water, it will stick so close to the bottom, that on taking the fish by the tail, one may lift up pail and all, though it hold several gallons of water. Great numbers of these fish are found along the coast of Greenland in the beginning of summer, where they resort to spawn. Their roe is remarkably large, and the Greenlanders boil it to a pulp for eating. They are extremely fat, but not admired in England, being both flabby and insipid.

THE UNCTUOUS SUCKER, OR SEA SNAIL,

TAKES its name from the soft and unctuous texture of its body, resembling the snail upon land. It is almost trans-

parent, and soon dissolves, and melts away. It is but a
little animal, being not above five inches long. The colour, when fresh taken, is of a pale brown, and the shape of
the body is round. It is taken in England, at the mouths
of rivers, four or five miles distant from the sea.

THE PIPE FISH.

THE body of the Pipe Fish, in the thickest part, is not
thicker than a swan quill, while it is above fifteen inches
long. Its general colour is an excellent olive brown,
marked with numbers of bluish lines, pointing from the
back to the belly. It is viviparous; for, on crushing one
that was just taken, hundreds of very minute young ones
were observed to crawl about.

THE HIPPOCAMPUS,

WHICH, from the form of its head, some call the Sea
Horse, never exceeds nine inches in length. It is about
as thick as a man's thumb; and the body is said, while
alive, to have hair on the fore part, which falls off when it
is dead. The snout is a sort of tube, with a hole at the
bottom, to which there is a cover, which the animal can
open and shut at pleasure. Behind the eyes there are
two fins which look like ears; and above them are two
holes, which serve for respiration. It, upon the whole,
more resembles a great caterpillar than a fish.

To these animals may be added the GALLEY FISH,
which Linnæus degrades into the insect tribe, under the
title of the Medusa. To the eye of an unmindful spectator
this fish seems a transparent bubble, swimming on the surface of the sea, or like a bladder variously and beautifully
painted with vivid colors, where red and violet predominate,
as variously opposed to the beams of the sun. It is, how-

ever, an actual fish; the body of which is composed of
cartilages, and a very thin skin filled with air, which thus
keep the animal floating on the surface, as the winds and
the waves happen to drive. Persons who happen to be
walking along the shore, often tread upon these animals;
and the bursting of their body yields a report as when one
treads upon the swim of a fish. It has eight broad feet
with which it swims, or which it expands to catch the air
as with a sail. It fastens itself to whatever it meets by
means of its legs, which have an adhesive quality. But
what is most remarkable in this extraordinary creature is
the violent pungency of the slimy substance with which
its legs are smeared. If the smallest quantity but touch
the skin, so caustic is its quality, that it burns it like hot
oil dropped on the part affected. The pain is worst in
the heat of the day, but ceases in the cool of the evening.

CHAP. VIII.

BONY FISHES.

THE third general division of fishes is into that of the Spinous, or Bony Kind. These are obviously distinguished from the rest, by having a complete bony covering to their gills; by their being furnished with no other method of breathing but gills only; by their bones, which are sharp and thorny; and their tails, which are placed in a situation perpendicular to the body.

The history of any one of this order very much resembles that of all the rest. They breathe air and water hrough the gills; they live by rapine, each devouring such animals as its mouth is capable of admitting; and they propagate, not by bringing forth their young alive, as in the cetaceous tribes, nor by distinct eggs, as in the generality of the cartilaginous tribes, but by spawn, or peas, as they are generally called, which they produce by hundreds of thousands.

The bones of this order of fishes, when examined but slightly, appear to be entirely solid; yet, when viewed more closely, every bone will be found hollow, and filled with a substance less rancid and oily than marrow. These bones are very numerous, and pointed; and, as in quadrupeds, are the props or stays to which the muscles are fixed, which move the different parts of the body.

The number of bones in all spinous fishes of the same kind is always the same. It is a vulgar way of speaking, to say, that fishes are, at some seasons, more bony than at others; but this scarce requires contradiction. It is true, indeed, that fish are at some seasons much fatter than at others; so that the quantity of the flesh being diminished, and that of the bones remaining the same, they appear to increase in number, as they actually bear a greater proportion.

As the spinous fishes partake less of the quadruped in their formation than any others, so they can bear to live out of their own element a shorter time. Some, indeed, are more vivacious in air than others; the eel will live several hours out of water; and the carp has been known to be fattened in a damp cellar. The method is, by placing it in a net well wrapped up in wet moss, the mouth only out, and then hung up in a vault. The fish is fed with white bread and milk, and the net now and then plunged into the water.

It is impossible to account for the different operations

of the same element upon animals that to appearance have the same conformation. To some fishes, bred in the sea, fresh water is immediate destruction; on the other hand, some fishes, that live in our lakes and ponds, cannot bear the salt water. This circumstance may possibly arise from the superior weight of the sea water. As, from the great quantity of salt dissolved in its composition, it is much heavier than fresh water, so it is probable it lies with greater force upon the organs of respiration, and gives them their proper and necessary play: on the other hand, those fish which are used only to fresh water, cannot bear the weight of the saline fluid, and expire in a manner suffocated in the grossness of the strange element. There are some tribes, however, that spend a part of their season in one, and a part in the other. Thus the salmon, the shad, the smelt, and the flounder, annually quit the ocean, and go up the rivers to deposit their spawn. This seems the most important business of their lives; and there is no danger which they will not encounter, even to the surmounting precipices, to find a proper place for the deposition of their future offspring. The salmon, upon these occasions, is seen to ascend rivers five hundred miles from the sea, and to brave, not only the dangers of various enemies, but also to spring up cataracts. As soon as they come to the bottom of the torrent, they seem disappointed to meet the obstruction, and swim some paces back; they then take a view of the danger that lies before them, survey it motionless for some minutes, advance, and again retreat; till at last, summoning up all their force, they take a leap from the bottom, their body straight, and strongly in motion; and thus most frequently clear every obstruction. It sometimes happens, however, that they want strength to make the leap; and then, in the fisheries, they are taken in their descent.

But the length of the voyage performed by these fishes is sport, if compared to what is annually undertaken by some tribes that constantly reside in the ocean. Of this kind are the cod, the haddock, the whiting, the mackerel, the tunny, the herring, and the pilchard.

The power of increasing in these animals exceeds our idea, as it would, in a very short time, outstrip all calculation: and a single herring, if suffered to multiply unmolested and undiminished for twenty years, would show a progeny greater in bulk than ten such globes as that we live upon. Although the usual way with spinous fishes is to produce by spawn, yet there are, some, such as the eel and the blenny, that are known to bring forth their young alive.

With respect to the growth of fishes, it is observed, that among carps particularly, the first year they grow to about the size of the leaf of a willow tree; at two years, they are about four inches long. They grow but one inch more the third season, which is five inches. Those of four years old are about six inches; and seven after the fifth. From that to eight years old they are found to be larger in proportion to the goodness of the pond, from eight to twelve inches. With regard to sea-fish, the fishermen assure us, that a fish must be six years old before it is fit to be served up to table. They instance it in the growth of a mackerel. They assure us that those of a year old are as large as one's finger; and those of two years, are about twice that length; at three and four years, they are that small kind of mackerel that have neither milts nor roes; and between five and six, they are those full grown fish that are served up to our tables. In the same manner, with regard to flat fishes, they tell us that the turbot and plaise at one year are about the size of a crown piece; the second year as large as the palm of one's hand; and, at the fifth and sixth year, they are large enough to be served up to table. Thus, it appears, that fishes are a considerable time in coming to their full growth, and that they are a long time the prey of others before it comes to their turn to be destroyers.*

The greediness with which sea fish devour the bait is prodigious, if compared with the manner they take it in fresh water. The lines of such fishermen as go off to sea, are coarse, thick, and clumsy, compared to what are used by those who fish at land. Their baits are seldom more than a piece of fish, or the flesh of some quadruped, stuck on the hook in a bungling manner; and scarce any art is employed to conceal the deception. But it is otherwise in fresh water; the lines must often be drawn to a hairlike fineness; they must be tinctured of the peculiar colour of the stream; the bait must be formed with the nicest art, and even, if possible, to exceed the perfection of nature: yet still the fishes approach it with diffidence, and often swim round it with disdain. The cod, on the banks of Newfoundland, the instant the hook, which is only baited with the guts of the animal last taken, is dropped into the water, darts to it at once, and the fishermen have but to pull up as fast as they throw down. But it is otherwise with those who fish in fresh waters, they must wait whole hours in fruitless expectation; and *the patience of a fisherman* is proverbial.

As fish are enemies to one another, so each species is infested with worms of different kinds, peculiar to itself. The great fishes abound with them; and the little ones are not entirely free. These troublesome vermin lodge themselves either in the jaws, and the intestines internally, or near the fins without. When fish are healthy and fat, they are not much annoyed by them; but in winter, when they are lean or sickly, they then suffer very much.

* Traite des Peches, par Monsieur Duhamel. Sec. iii. p. 100.

Nor does the reputed longevity of this class secure them from their peculiar disorders. They are not only affected by too much cold, but there are frequently certain dispositions of the element in which they reside, unfavourable to their health and propagation. Some ponds they will not breed in, however artfully disposed for supplying them with fresh recruits of water, as well as provision. In some seasons also they are found to feel epidemic disorders, and are seen dead by the water side, without any apparent cause.

The fact of some fishes in warm climates being poisonous when eaten, cannot be doubted. There is a paper in the London Philosophical Transactions, giving an account of the poisonous qualities of those found at New Providence, one of the Bahama islands. The author there assures us, that the greatest part of the fish of that dreary coast are all of a deadly nature; their smallest effects being to bring on a terrible pain in the joints, which, if terminating favourably, leaves the patient without any appetite for several days after. It is not those of the most deformed figure, or the most frightful to look at, that are alone to be dreaded; all kinds, at different times, are alike dangerous; and the same species which has this day served for nourishment, is the next, if tried, found to be fatal.

As this order of fishes is extremely numerous, various modes of classing them have been invented by different naturalists. The simplest is that of Linnæus, who ranks them in four divisions, according to the position of the fins.

The first division is what that celebrated naturalist terms APODAL. This includes the most imperfect of the order, viz. *those which want the ventral or belly fins* (as the wolf fish), and consists of the following genera.

THE EEL

Is the first genus of this division, and includes several species.

The COMMON EEL is a very singular fish in several things that relate to its natural history, and in some respects borders on the nature of the reptile tribe.

It is known to quit its element, and, during night, to wander along the meadows, not only for change of habitation, but also for the sake of prey, feeding on the snails it finds in its passage.

During winter, it beds itself deep in the mud, and continues in a state of rest like the serpent kind. It is very impatient of cold, and will eagerly take shelter in a wisp of straw flung into a pond in severe weather, which has sometimes been practised as a method of taking them. Albertus goes so far as to say, that he has known Eels to shelter in a hay-rick, yet all perished through excess of cold.

It has been observed, that in the river Nen in England, there is a variety of small Eel, with a lesser head and narrower mouth than the common kind; that it is found in clusters in the bottom of the river, and is called the Bed Eel; these are sometimes roused up by violent floods, and are never found at that time with meat in their stomachs. This bears such an analogy with the clustering of blind worms in their quiescent state, that we cannot but consider it as a further proof of partial agreement in the nature of the two genera.

The ancients adopted a most wild opinion about the generation of these fish, believing them to be either created from the mud, or that the scrapings of their bodies which they left on the stones, were animated, and became young Eels. Some moderns gave into these opinions, and into others that were equally extravagant. They could not account for the appearance of these fish in ponds that were never stocked with them, and were even so remote as to make their being met in such places a phenomenon that they could not solve. But there is much reason to believe, that many waters are supplied with these fish by the aquatic fowl of prey, in the same manner as vegetation is spread by many of the land birds, either by being dropped as they carry them to feed their young, or by passing quick through their bodies, as is the case with herons; and such may be the occasion of the appearance of these fish in places where they were never seen before. Or in some of the nocturnal wanderings, already mentioned, Eels may have strayed into those new places of abode. As to their immediate generation, it has been sufficently proved to be effected in the ordinary course of nature, and that they are viviparous.

They are extremely voracious, and very destructive to the fry of fish.

No fish lives so long out of water as the Eel; it is extremely tenacious of life, and its parts will move a considerable time after they are flayed and cut in pieces.

The Eel is placed by Linnæus in the genus of *murœna*, his first of the apodal fish, or those which want the ventral fins.

The eyes are placed not remote from the end of the nose : the irides are tinged with red : the under jaw is longer than the upper ; the teeth are small, sharp, and numerous ; beneath each eye is a minute orifice; at the end of the nose two others, small and tubular. This fish is furnished with a pair of pectoral fins, rounded at their ends. Another narrow fin on the back, uniting with that of the tail ; and the anal fin joins it in the same manner beneath. Behind the pectoral fins is the orifice to the gills, which are concealed in the skin.

Eels vary much in their colours, from a sooty hue to a light olive green ; and those which are called Silver Eels have their bellies white, and a remarkable clearness throughout.

Besides these, there is another variety of this fish, known in the Thames by the name of Grigs, and about Oxford by that of Grigs or Gluts. These are scarce ever seen near Oxford in the winter, but appear in spring, and bite readily at the bait, which common Eels in that neighbourhood will not. They have a larger head, a blunter nose, thicker skin, and less fat, than the common sort; neither are they so much esteemed, nor do they often exceed three or four pounds in weight.

Common Eels grow to a large size, sometimes so great as to weigh fifteen or twenty pounds, but that is extremely rare. As to instances brought by Dale and others, of these fish increasing to a superior magnitude, we have much reason to suspect them to have been congers, since the enormous fish they describe have all been taken at the mouth of the Thames or Medway, in England.

The Eel is the most universal of fish, yet is scarce ever found in the Danube, though it is very common in the lakes and rivers of Upper Austria.

The Romans held this fish very cheap, probably from its likeness to a snake.

"For you is kept a sink-fed snakelike eel."
JUVENAL, Sat. v.

On the contrary, the luxurious Sybarites were so fond of these fish, as to exempt from every kind of tribute the persons who sold them.

The CONGER EEL grows to an immense size, and its fierceness is equal to its magnitude; they have been taken ten feet and a half long, and eighteen inches in circumference in the thickest part. They differ from the common Eel not only in their size, but in being of a darker colour, and in the form of the lower jaw, which is shorter than the upper. They are extremely voracious, and prey upon other fish, particularly upon crabs, when they have cast their shell. The fishermen are very fearful of the large congers, lest they should endanger their legs by clinging round them; they therefore kill them as soon as possible, by striking them on the navel. In April, 1808, one was taken at Yarmouth, Eng. which knocked down its captor before it could be secured. On the coast of Cornwall these fish constitute a considerable article of commerce, where they are salted and dried, and afterwards ground to powder, which is purchased by the Spaniards for the purpose of thickening their soups.

The NETTED EEL. The head and mouth of this beautiful species is small, without barbles. The eyes are near the upper lip, of a blue and white colour. The teeth stand apart from each other, and those in front are the longest. The body is variegated with black and white spots like network, and the dorsal fin extends the whole length of the back.

This species is from two to three feet in length, and was

F3

found near the coast of Tranquebar; but little is known of its natural history.

The CORDATED EEL. This fish is a native of the West Indian seas; its whole length is about thirty-two inches, of which the process at the end of the tail measures twenty-two; the body of a rich silvery hue, the flexible part belonging to the snout brown, the fins and caudal process a paler brown. The snout is connected to the back part of the head by a flexible leathery duplicature which permits it to be extended so that the mouth points upwards, or to fall back so as to be received into a sort of case formed by the upper part of the head; below the head, on each side, is a considerable compressed semicirular space, the forepart of which is bounded by the gill-cover, which seems to consist of a moderately strong membrane; the body gradually diminishes as it approaches the tail, which terminates in a process or string of enormous length, ending in a very fine point; the pectoral fins are small, situate behind the cavity on each side the thorax; the caudal fin has five spinous rays.

But the most extraordinary fish of this kind is the ELECTRICAL EEL, or GYMNOTUS. Of the Gymnotus tribe some

of the species inhabit the fresh water, and others the ocean, and with the exception of three of them all are confined to the continent of America. The species which is the subject of the engraving is common in South America. It is from three to five feet in length, and ten or twelve inches in circumference in the broadest part of the body; and has the capability of swimming backward as well as forward. Their colour is an olive green, and the head yellow mingled with red. The head is flat, and the mouth wide and toothless. From the point of its tail to within six inches of its head extends a fin about two inches deep, and which is an inch thick at its junction with the body. As there are several annular divisions, or rather rugæ of the skin, across the body, it would seem that the fish partakes of the vermicular nature, and can contract or dilate itself at pleasure.

The electrical shock is conveyed either through the hand, or any metallic conductor which touches the fish; and a stroke of one of the largest kind, if properly applied, would prove instant death to even the human species. This extraordinary power is given to this fish, not only for defence, but subsistence. For whenever small fishes or worms are thrown into the water, they are first struck dead by the electric power of the animal, and afterwards swallowed by him.

M. Humboldt gives an interesting account of the manner in which they catch these creatures, by what they call " fishing with horses." After having scoured the savannah, and caught about thirty wild horses and mules, they drove them into the pool in which were the electrical Eels. " The extraordinary noise caused by the horses' hoofs makes the fish issue from the mud, and excites them to combat. These yellowish and livid Eels, resembling larger aquatic serpents, swim on the surface of the water, and crowd under the bellies of the horses and mules. A contest between animals of so different an organization furnishes a very striking spectacle. The Indians, provided with harpoons and long slender reeds, surround the pool closely; and some climb upon the trees, the branches of

which extend horizontally over the surface of the water.
By their wild cries, and the length of their reeds, they
prevent the horses from running away, and reaching the
bank of the pool.   The Eels, stunned by the noise, defend
themselves by the repeated discharge of their electric bat-
teries.   During a long time they seem to prove victorious.
Several horses sink beneath the violence of the invisible
strokes, which they receive from all sides in organs the most
essential to life ; and stunned by the force and frequency of
the shocks, disappear under the water. Others panting, with
mane erect, and haggard eyes expressing anguish, raise
themselves, and endeavour to flee from the storm by which
they are overtaken.   They are driven back by the Indians
into the middle of the water; but a small number succeed
in eluding the active vigilance of the fishermen.   These
regain the shore, stumbling at every step, and stretch
themselves on the sand, exhausted with fatigue, and
their limbs benumbed by the electric shocks of the Gymnoti.

"In less than five minutes two horses were drowned.
The Eel being five feet long, and pressing itself against
the belly of the horse, makes a discharge along the whole
extent of its electric organ.   It attacks at once the heart,
the intestines, and the *plexus cœliacus* of abdominal nerves.
It is natural that the effect felt by the horses should be
more powerful than that produced upon man by the touch
of the same fish at only one of his extremities.   The
horses are probably not killed, but only stunned.   They
are drowned from the impossibility of rising amid the pro-
longed struggle between the other horses and the Eels.

"We had little doubt that the fishing would terminate
by killing successively all the animals engaged ; but by
degrees the impetuosity of this unequal contest diminished,
and the wearied Gymnoti dispersed.   They require a long

rest,* and abundant nourishment, to repair what they have lost of galvanic force. The mules and horses appear less frightened; their manes are no longer bristled, and their eyes express less dread. The Gymnoti approach timidly the edge of the marsh, where they are taken by means of small harpoons fastened to long cords. When the cords are very dry, the Indians feel no shock on raising the fish into the air. In a few minutes we had five large Eels, the greater part of which were but slightly wounded."

THE LAUNCE, OR SAND EEL,

Is known by a body slender and roundish; the head terminated by a beak; the teeth of a hairlike fineness; the fin covering the gills with seven rays. It grows to the length of nine or ten inches, and is found in most of our sandy shores during the summer months. It conceals itself among the sand, whence, during flood tide, they are rooted up, and devoured by the porpesses; and on the recess of the tide they are drawn out with a hook by the fishermen. They are commonly made use of as a bait for other fish, but they are also very delicate eating.

THE WOLF FISH

Has the body roundish and slender; the head large and blunt; the fore teeth, above and below, conical; the grinding teeth, and those in the palate, round; the fin covering the gills has seven rays.

This animal seems to be confined to the northern seas, and sometimes is found near the coasts of Scotland. It grows to a very large size, being frequently taken of the length of seven feet, and even more. It is a most raven-

---

* The Indians assured us, that when the horses are made to run two days successively into the same pool, none are killed the second day.

ous and fierce fish, and when taken, fastens upon every thing within its reach. It is said even to bite so hard, that

it will seize upon an anchor, and leave the marks of its teeth on it. It feeds almost entirely on shell-fish, the hardest of which it easily crushes with its jaws. It has so formidable and disagreeable an appearance, that it is only eaten by the fishermen, who, however, prefer it to halibut.

THE SWORD FISH

Is very common in the Mediterranean, and is much esteemed for food by the Sicilians, who consider it as equal to the sturgeon. It is also found on the coasts of America. It grows to a very large size, upwards of twenty feet in length. It is of a long and rounded body, largest near the head, and tapering by degrees to the tail. The skin is rough, the back black, and the belly white. It has one fin on the back, running almost its whole length. It has one dair of fins also at the gills. But the most remarkable part of this fish is the snout, which, in the upper jaw, runs out in the figure of a sword, sometimes to the length of three feet, and is of a substance like a coarse kind of ivory. The under jaw is much shorter.

The Sword Fish has wonderful strength. The Leopard man of war was struck by one of them; and though the animal was following the ship, and consequently gave the blow with less force than it otherwise would have done, yet the sword penetrated nearly a quarter of a yard through the sheathing and timber, and was broken off by the shock. Eight or nine strokes from a hammer weighing a quarter of a hundred weight would be required to drive an iron pin the same depth into wood. In the British Museum there is also a plank of a ship, through which a fish impelled the whole length of his sword; not, however, without losing his life by the effort.

The Sword Fish has an antipathy to the whale, and no sooner meets than he assails him. Two will sometimes combine in the attack. The whale can defend himself only with his tail, which the activity of his adversary generally enables him to evade. The whale dives in vain, for he is pursued by his pertinacious tormentor, and he is at length compelled to take flight.

There is another species of Sword Fish, called the BROAD-FINNED. Both are insatiably voracious.

The second division consists of the JUGULAR FISHES, or those *which have ventral fins before the pectoral*, or nearer to the gills.

THE DRAGONET

Is the first genus which naturalists have remarked in this division. Its general characters are the upper lip doubled, the eyes very near each other, two breathing apertures on the hind part of the head, and the first rays of the dorsal fin extremely long. There are two species described by Mr. Pennant.

The GEMMEOUS DRAGONET, which is about ten or twelve, inches long, with a large head, and a body slender, round, and smooth. The colours of this fish are extremely beautiful; when it is just taken, they are yellow, blue, and white. The blue has all the splendour of the most beautiful gems. The throat is black; and the membranes of the fins are very thin and delicate. The old English writers have called this fish the yellow gurnard, but in reality it has no one character of that genus.

The SORDID DRAGONET resembles the preceding, but its first dorsal fin is not so long, nor are its colours so brilliant and lively.

THE WEEVER

Is known sometimes to grow to the length of twelve inches, though it is commonly found much smaller. The lower jaw slopes down very abruptly, and its back is armed with strong spines. It buries itself in the sand, leaving only its nose out, and when trod upon strikes forcibly with its spines, which are said to be venomous, though probably the pain and inflammation attending the wounds which it inflicts depend on the habit of the person or the part which is struck. It is good food.

THE COD

Is a most extensive genus, including a number of well known and useful fishes. The general characters are a smooth head, the fin that covers the gills consisting of seven rays, all the fins covered with a common skin, the ventral fins slender, and ending in a point. It has teeth in the jaws, and a series of small teeth closely set together in the palate. Most of the species have also the chin bearded.

The COMMON COD is short in proportion to its bulk;

the belly is large and prominent; its eyes are large; and at the end of the lower jaw is a small beard. It is ash-coloured, spotted with yellow, and the belly white; on the back are three soft fins. It is one of the most prolific of fish.

There are also the THREE-BEARDED and FIVE-BEARD-ED CODS, both of which differ from the common sort, not only in this character, but in having only two back fins, the latter very long.

The Cod seems to be the foremost of the wandering tribe of fishes, and is only found in the northern part of the world. Their principal food consists of the smaller species of fish, worms, shell-fish, and crabs; and their stomachs are capable of dissolving the major part of the shells which they swallow. They grow to a great size. The largest that ever was seen was taken at Scarborough, England, in 1775. It weighed seventy-eight pounds, and was five feet eight inches long. This animal's chief place of resort is on the banks of Newfoundland, and the other sand banks that lie off Cape Breton. That extensive flat seems to be no other than the broad top of a sea mountain, extending for above five hundred miles long, and surrounded with a deeper sea. Hither the Cod annually repair in numbers, beyond the power of calculation, to feed on the quantity of worms that are to be found there in the sandy bottom. Here they are taken in such quantities, that they supply Europe and America with a considerable share of provision. The English have stages erected all along the shore for salting and drying them; and the fishermen, who take them with the hook and line, which is their method, draw them in as fast as they can throw out. An expert hand will sometimes capture four hundred in a day. This immense capture, however, makes but a very small diminution, when compared to their numbers; and when their provision there is exhausted, or the season for propagation returns, they go off to the polar seas, where they deposit their spawn. Previous to the discovery of Newfoundland, the principal fisheries for Cod were in the Iceland seas, and off the western isles of Scotland.*

THE HADDOCK

Is a well known fish of this genus, which much resembles the cod, but is smaller; it is also distinguished by a black mark on each side beyond the gills, which superstition ascribes to the impression which St. Peter left with his finger and thumb, when he took the tribute money out of the fish's mouth, which tradition would have us believe to have been of this species.

THE WHITING POUT

Is another fish of the same kind, which in size seldom exceeds a foot. The back is much arched; the scales larger than that of the cod; and on each side of the jaw are seven or eight punctures.

* Considerable quantities of Cod fish are taken every year at Nantucket Shoals, and at the Isles of Shoals, on the coasts of New England.

THE RIB

Grows also to the length of a foot, and the sides are finely tinged with gold.

THE POOR

Is the only species of Cod found in the Mediterranean; it is not more than six inches long.

THE COAL FISH

Takes its name from the black colour it sometimes assumes. It grows to the length of two feet and a half, and is of a more elegant shape than the Cod. The flesh is little esteemed when fresh, but is commonly salted and dried for sale. The fry of this fish, however, is called *Parr*, and is esteemed good food.

THE POLLOCK

Does not grow to a very large size, but is a very good eating fish. The first back fin has eleven rays, the middle nineteen, the last sixteen. The tail is a little forked; the colour of the back is dusky, in some inclining to green the belly is white.

THE WHITING

Is a fish of an elegant form, and the most delicate food of all the genus. The first back fin has fifteen rays, the second eighteen, the third twenty. The back is a pale brown, and the belly silvery white. It seldom exceeds twelve inches in length.

This fish is found in the Baltic and North seas, though not numerous in the former; but they are plentiful on the coasts of Holland, France, and England, where they are reckoned the most delicate and wholesome species of the genus. Their flesh is so easily digestible, that it is prescribed to persons the powers of whose stomachs are impaired. They attain the length of a foot, sometimes one and a half, rarely two; but on the Doggerbank they are caught of the weight of from four to eight pounds. They live at the bottom of the sea, feeding on little crabs, worms, and young fry, particularly of sprats and herrings; which therefore are the usual baits. They are caught usually with a ground-line, sixty-four fathoms long, with from a hundred to two hundred hooks. One vessel throws out about twenty of these lines, armed with four thousand hooks, and they need only lie about two or three hours. The greatest fishery for Whitings is carried on by the French from December to February; by the English and Dutch in the spring. They appear in such quantities on the English coasts, as to form shoals of three miles long and a mile and a half wide; and, as they are caught in too great numbers to be eaten fresh, they salt them, by which, however, they lose the delicacy of their taste, and are then called buckthorn; they are often used in this state as ships' stores. As they eagerly pursue the herrings, they are often taken in the same nets.

The seven last species have three back fins; the HAKE, the LING, and the BURBOT, have only two; and the TORSK has only one.

THE BLENNY

HAS the body oblong; the head obtuse; the teeth a single range; the fin covering the gills with six spines; the ventral fins have two small blunt bones in each. It has one dorsal fin which is prickly, and several of the species are crested, or have a small fin like a crest upon their heads. It is a small fish, measuring from five to seven inches, and is found among stones upon rocky coasts, and sometimes in the mouths of rivers. One species of this fish is viviparous, and brings forth two or three hundred at a time. These are very common at the mouth of the Eske, at Whitby, in Yorkshire, England.

THE BUTTERFLY-FISH.

This fish, which is a species of Blenny, has a long, large head, narrowed at the sides; and prominent eyes of considerable magnitude, with black pupil, and orange-coloured iris; the mouth is large; the jaws are of equal length, with a row of very narrow teeth standing close to each other; the tongue is broad, but short. The gills are wide, the cheeks large, and of a silver colour. The back is round, and of a dark green; the belly is short but broad. The ground colour of the fish is a dirty green, with brown spots; there are some, however, whose principal colour is a clear blue. The dorsal fin is spotted with black. This fish inhabits the Mediterranean Sea; and at Marseilles, Sardinia, and Venice, is common in the markets with other small fish. It attains to the length of six or eight inches; lives near the shore, among the rocks and weeds; and feeds on crabs and small shell-fish. Its scales are hardly visible. Some naturalists describe this fish as having two dorsal fins; while others say it has but one. This contradiction must arise from these fins being sometimes united by a membrane, and sometimes not.

THE STAR GAZER, OR URANOSCOPUS,

Has a large quadrangular head, covered with a rough helmet, ending in two spines above, and five smaller ones below. The mouth opens upwards, and when the lower jaw is removed, the tongue appears, which is thick, short, and strong, and full of small teeth. Inside of the lower jaw, there is a membrane terminating by a long filament; the fish, opening its mouth, sets this in motion, which attracts little fish, who endeavour to seize it, and are instantly devoured. There are two barbles from each lip, which serve for the same purpose; and this fish often conceals itself among the sea-weed, leaving only the barbles visible when it is watching for its prey. In the upper jaw there are two oval apertures, and several little barbles at the lower; and near each eye is a round aperture. The eyes lie quite at the top of the head, very close together, and prominent, as if starting upwards; the pupil is black, the iris yellow. The Star Gazer inhabits the Mediterranean, in deep places near the shore. It seldom exceeds a foot in length.

The third division is called the THORACIC, or *those fishes which have the belly fins immediately under the pectoral.*

THE GOBY, OR ROCK-FISH,

Is not above six inches long. The body is soft, slippery, and slender; the head large; the cheeks inflated. It has two back fins; and the ventral fins coalesce, and form a sort of funnel, by which these fish fix themselves immoveably to the rocks.

THE REMORA, OR SUCKING-FISH,

WHICH has been already in part described, appears to belong to this genus. In shape it resembles a herring, but on the head has an apparatus for fixing itself to a ship, or to the body of another fish. It is an inhabitant of the Indian Ocean.

THE BULL-HEAD

Is a well known genus, including several species, all of which have a large head armed with spines.

THE RIVER BULL-HEAD, OR MILLER'S THUMB,

Is very common in all our clear lakes. It rarely exceeds three inches in length, and is easily distinguished by a broad flat head, excellently adapted for insinuating itself under stones. It is of a dusky colour, mixed with dirty yellow, and spotted with white, black, and brown; and has two back fins.

THE POGGE, OR ARMED BULL-HEAD,

Is found on most of the European coasts, and is distinguished by its large bony head, which is armed at the nose with four short upright spines, and by a number of white beards at the throat. It is about five inches long.

But the most formidable of this genus is the FATHER LASHER, or SEA SCORPION. It is about eight or nine inches long. The nose, the top of the head, and the back fins are armed with strong sharp spines. It is exceedingly common in the Newfoundland seas, and makes a principal article of food in Greenland.

THE FLYING SCORPÆNA.

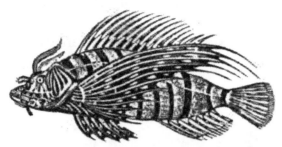

This fish has a truncated head, which is broad in front, compressed at the sides, and furnished with pretty large spines and fringed barbles; the longest of which are over the eyes, and the broadest near the corners of the mouth. On the body and head are several brown stripes, with yellow and white streaks alternately shining between. The mouth is large; the jaws are of equal length, and armed with a great number of little sharp teeth. The tongue is loose, thin, and pointed at the end; the lips are also moveable; the upper lip is composed of two bones, which form a furrow in the middle where they join. The nostrils are single, and lie midway between the mouth and eyes. The eyes have a black pupil, and a white iris with blue and black rays. The gilt covert terminates in a sharp angle, and is furnished with very minute scales; the aperture is wide, and the bronchial membrane is in great part naked. The scales on the body are small, and lie one over the other like tiles on a house. The lateral line consists of little risings and white points. The rays of the pectoral fins are simple, and the membrane has a violet ground with white dots. These large fins probably enable the fish to dart out of the water when pursued by an enemy. The first twelve rays of the dorsal fin are spiny, spotted brown and yellow; united below by a dark brown membrane, and at liberty above; the last twelve rays, as well as those of the anal and tail fins, are divided at the ends, and spotted black and yellow. The ventral fins are violet, with white dots. The first ray is hard. The skin is like parchment.

This variegated fish is found in the rivers at Amboyna and Japan; but even there it is uncommon. It is known also at Tranquebar. The flesh is white, firm, and well tasted, like our perch, but the fish does not grow so large. It is of the voracious kind, feeding on the young of other species; entire fishes of two inches and a half long have been found in the stomach.

THE DOREE, OR JOHN DOREE,

Is mentioned in the writings of Ovid and Pliny. It is almost equally famous in the legends of superstition with the haddock, and is its rival in the honour of being the fish from which St. Peter took the tribute money, leaving on its sides the mark of his finger and thumb. The modern Greeks call it the fish of St. Christopher, from a legend, which relates that it was trodden on by that saint, as he was carrying his divine burthen across an arm of the sea.

The form of this fish is very disgusting. Its body is oval, and much compressed at the sides. Its snout is long, and its mouth is wide. The first back fin consists of ten spiny rays, with long filaments; the second of twenty-four soft rays. The tail is round at the end. The colour of the body is olive, varied with light blue and white; while living it has the appearance of gilding, whence its name *Dorée* (gilt). It is found in the North Sea, the British Channel, the Atlantic, and the Mediterranean.

Such is the unpleasant form of this fish, that it was long before it attracted the notice of the epicure. Mr. Pennant observes, indeed, that to the celebrated actor, Mr. Quin, it is chiefly indebted for its reputation. Quin is said to have travelled to Plymouth that he might eat this fish in perfection; and to have exclaimed that if, in angling for him, his Satanic Majesty were to bait his hook with a Doree, he could not refrain from biting.

THE OPAH

Is another of this genus, which sometimes arrives at an immense size. One was caught at Torbay in 1772, which weighed a hundred and forty pounds. It was in length four feet and a half; in breadth two feet and a quarter, though the greatest thickness was only four inches. The general colour was a transparent scarlet varnish, spangled with silver spots of various sizes. The mouth of this fish is exceedingly small for its size.

THE FLOUNDER

Is a very extensive genus, including those innumerable species which are known by the common term flat fish, and which are distinguished from all others by one invariable characteristic, viz. that of having both the eyes on the same side of the head.

THE HALIBUT

Is by much the largest of the genus, weighing commonly from one hundred to three hundred pounds. The Halibut is the most voracious of fishes, and has been known to swallow even the lead which seamen make use of for the purpose of sounding the depth. Its back is a dusky colour; its belly pure white. The flesh is very coarse and indifferent food. It is the narrowest fish in proportion to its length of any of this genus, except the sole.

THE PLAISE

Is sometimes known to weigh fifteen pounds. It is easily distinguished by the upper part of the body, which is dusky, being marked with large orange-coloured spots.

THE FLOUNDER, OR FLUKE,

MAY be easily known from every other fish of this genus, by a row of sharp small spines, which surround its upper sides, and are just placed where the fins join to the body. It differs principally from the plaise in wanting the row of six tubercles behind the left eye. It is found in most of the European rivers, up which it ascends even beyond the influence of the tide. The back is of a pale brown, sometimes marked with a few obscure spots.

THE DAB

Is found often along with the flounder, but is less common. It is smaller than either the plaise or the flounder, but is

LIKE some others of the flat fish, grows to a great size. It has occasionally been known to weigh from twenty-five to thirty pounds. In its general form it is somewhat square. Flat fish swim sideways, on which account they are styled pleuronectes by Linnæus. The eyes of all of them are situated on one side of the head, those of the Turbot on the left; and it is a curious circumstance that, while the under parts of their body are of a brilliant white, the upper parts are so coloured and speckled as, when they are half immersed in the sand or mud, to render them imperceptible. Of this resemblance they are so conscious, that, whenever they find themselves in danger, they sink into the mud, and continue perfectly motionless. This is a circumstance so well known to fishermen, that within their palings on the strand they are often under the necessity of tracing furrows with a kind of iron sickle, to detect by the touch what they are not otherwise able to distinguish. But the Turbot does not thus hide itself for security alone. It resorts to this stratagem as an ambush for obtaining its prey,

---

* The Turbot we believe is not found in America. A fish much resembling it has been occasionally taken off Boston harbour.

whence it pounces forth on the smaller kinds of fish that incautiously approach it.

The finest Turbot in the world are found off the northern shore of England, and some parts of the Dutch coast. The manner of fishing for them off the Yorkshire coast is as follows: three men go out in each of the boats, each man provided with three lines; every one of which is furnished with two hundred and eighty hooks, placed exactly six feet two inches asunder. These are coiled on an oblong piece of wickerwork, with the hooks baited, and placed very regularly in the centre of the coil. When they are used, the nine are generally fastened together so as to form one line with above two thousand hooks, and extending near three miles in length. This is always laid across the current. An anchor and buoy are fixed at the end of each man's line. The boats for this purpose are each about a ton burthen; somewhat more than twenty feet in length, and about five feet in width.

The general bait used for taking Turbots is fresh herring cut into proper sized pieces, at which they bite most readily; they are also partial to the smaller lampreys, pieces of haddocks, sandworms, muscles, and limpets; and when none of these are to be had, the fishermen use bullock's liver. They are so extremely delicate in the choice of their baits, as not to touch a piece of herring or haddock that has been twelve hours out of the sea; nor will they touch any bait that has been bitten by another fish.

THE SOLE.

THIS well known and delicious fish is remarkable for one extraordinary circumstance; they have been known to feed on shell-fish, although they are furnished with no apparatus whatever in their mouth for reducing them to a state calculated for digestion. The stomach, however, has a dissolvent power, which makes up for the want of a masticating apparatus. But the most usual food for Soles is the spawn and young of other fish.

These fish are found on all the British coasts; but those off the western shores are much superior in size to what are taken in the north, since they are sometimes found of the weight of six or seven pounds. The principal fishery for them is in Torbay, Eng. In the winter they usually retire into deep water: but frequent the sea-shores and the mouths of rivers at the approach of spring. On the sand-banks out at sea, they are caught in trawl nets, and on the shore they are taken in seine nets. The Sole will keep longer sweet out of water than almost any other fish, and is even much better for being kept a while.

THE SMOOTH SOLE, OR LANTERN FISH,

Is almost peculiar to the Cornish coast; and is thin, white, and nearly pellucid.

THE PEARL

RESEMBLES the turbot, but is inferior food; its back is of a deep brown, with spots of a dirty yellow.

The WHIFF resembles the halibut, but is smaller.

THE GILT-HEAD

TAKES its name from its predominant colour, the forehead and sides resembling gold, though the latter are tinged with brown. It has but one back fin, which reaches the whole length of the body. In form it in some degree resembles the bream. It is found in deep waters, on bold rocky shores; it subsists chiefly on shell-fish, and some of the species grow to the weight of ten pounds.

Besides the LUNATED, which is the most common, and takes its name from a semilunar gold spot under the eyes, there are the RED, and the TOOTHED or STREAKED GILT-HEADS, the last of which is distinguished by two canine teeth on each side.

THE STREAKED GILT-HEAD.

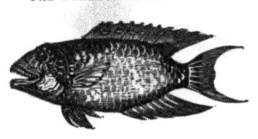

THE head is compressed, and scaleless as far as the eyes; the mouth large; the jaws are of equal length, with two strong canine teeth in each; the back teeth are flat, like those of quadrupeds. The nostrils are single, and near the eyes, which are small, with a black pupil and blue iris. The body is yellow, with six or seven brown transverse stripes. This fish is found on the shores of Japan and the Red Sea.

There is a fish in some degree resembling the preceding, which is called by naturalists, by way of eminence, the DORADO, but which the sailors erroneously term the dolphin; it is chiefly found in the tropical climates; and is at once the most active and the most beautiful of the finny race. It is about six feet long; the back all over enamelled with spots of a bluish green and silver;

the tail and fins of a gold colour; and all have a brilliancy
of tint, that nothing but nature's pencil can attain to : the
eyes are placed on each side of the head, large and beautiful, surrounded with circles of shining gold. In the seas
where they are found, these fish are always in motion, and
play round ships in full sail, with ease and security : for
ever either pursuing or pursued, they are seen continually in a state of warfare ; either defending themselves against
the shark, or darting after the smaller fishes.

THE FLYING FISH.

ABOVE all others, the Flying-fish most abounds in these
seas ; and as it is a small animal, seldom growing above
the size of a herring, it is chiefly sought by the dorado.
Nature has furnished each respectively with the powers of
pursuit and evasion. The dorado being above six feet long,
yet not thicker than a salmon, and furnished with a full com-
plement of fins, cuts its way through the water, with amazing rapidity ; on the other hand, the Flying-fish is furnished with two pair of fins, longer than the body, and these
also moved by a stronger set of muscles than any other.
This equality of power seems to furnish one of the most
entertaining spectacles those seas can exhibit. The efforts to seize on the one side, and the arts of escaping on
the other, are perfectly amusing. The dorado is seen,
upon this occasion, darting after its prey, which will
not leave the water, while it has the advantage of swim-

ming, in the beginning of the chase. But, like a hunted hare, being tired at last, it then has recourse to another expedient for safety, by flight. The long fins,
which began to grow useless in the water, are now
exerted in a different manner and different direction
to that in which they were employed in swimming: by
this means the timid little animal rises from the water,
and flutters over its surface, for two or three hundred
yards, till the muscles employed in moving the wings,
are enfeebled by that particular manner of exertion. By
this time, however, they have acquired a fresh power of
renewing their efforts in the water, and the animal is
capable of proceeding with some velocity by swimming:
still, however, the active enemy keeps it in view, and drives
it again from the deep; till at length, the poor little
creature is seen to dart to shorter distances, to flutter with
greater effort, and to drop down at last into the mouth of
its fierce pursuer. But not the dorado alone, all animated
nature seems combined against this little fish, which seems
possessed of double powers, only to be subject to greater
dangers: for, though it should escape from its enemies
of the deep, yet the tropic bird, and the albatross, are forever upon the wing to seize it. Thus pursued in either
element, it sometimes seeks refuge with a new enemy;
and it is not unfrequent for whole shoals of them to fall
on shipboard, where they furnish man with an object of
useless curiosity.

THE WRASSE

INCLUDES several species, the most common of which is
the ANCIENT WRASSE, or OLD WIFE. It is of a clumsy
shape, not unlike a carp, and covered with large scales;
it has one large back fin, which consists of sixteen sharp
spiny rays, and nine soft ones. The tail consists of four-

teen soft branching rays, and is rounded at the end. They
vary greatly in colour, some being of a dirty red, and
others beautifully striped. They are generally found in
deep water, adjacent to the rocks, and feed upon shell-
fish. They grow to the weight of four or five pounds.

Besides these species, Mr. Pennant has enumerated
the BALLAN, the BIMACULATED, TRIMACULATED, STRI-
PED, and GIBBOUS WRASSE, the GOLDSINNY, the SCOM-
BER, and the COOK.

THE PERCH

OF Aristotle and Ausonius is the same with that of the
moderns. That mentioned by Oppian, Pliny, and Athe-
næus, is a sea-fish, probably of the Labrus or Sparus kind,
being enumerated by them among some congenerous species. Our Perch was much esteemed by the Romans.
Nor is it less admired at present, as a firm and delicate
fish; the Dutch, indeed, are particularly fond of it when
made into a dish called water souchy.

It is a gregarious fish, and loves deep holes and gentle
streams. It is a most voracious fish, and eager biter; if
the angler meets with a shoal of them, he is sure of taking
every one. It is a common notion that the pike will not
attack this fish, being fearful of the spiny fins which the
Perch erects on the approach of the former. This may be
true in respect to large fish; but it is well known the
small ones are the most tempting bait that can be laid for
the pike.

The Perch is a fish very tenacious of life: we have known them carried near sixty miles in dry straw, and yet survive the journey. These fish seldom grow to a large size: we once heard of one that was taken in the Serpentine river, Hyde Park, London, that weighed nine pounds; but that is very uncommon.

The body is deep; the scales very rough; the back much arched; side-line near the back. The irides golden; the teeth small, disposed in the jaws, and on the roof of the mouth; the edges of the covers of the gills serrated; on the lower end of the largest is a sharp spine. The first dorsal fin consists of fourteen strong spiny rays; the second of sixteen soft ones; the pectoral fins are transparent, and consist of fourteen rays; the ventral of six; the anal of eleven. The tail is a little forked. The colours are beautiful; the back and part of the sides being of a deep green, marked with five broad black bars pointing downwards; the belly is white, tinged with red; the ventral fins of a rich scarlet; the anal fins and tail of the same colour, but rather paler.

In a lake called Llyn Raithlyn, in Merionethshire in Wales, is a very singular variety of Perch; the back is quite hunched, and the lower part of the back-bone, next the tail, strangely distorted; in colour, and in other respects, it resembles the common kind, which are as numerous in the lake as these deformed fish. They are not peculiar to this water; for Linnæus takes notice of a similar variety found at Fahlun, in his own country. We have also heard that it is to be met with in the Thames, near Marlow in England.

THE BASSE

Is a larger and coarser kind of perch, which sometimes grows to the weight of fifteen pounds. It is, however, of rather a longer make, more resembling that of a salmon. The back is dusky, tinged with blue, and the belly white. The Sea Perch grows to about a foot long. The head is large and deformed, and covered with sharp spines. The colour is red, with a black spot on the covers of the gill and some transverse dusky lines on the sides.

THE RUFFE

Is a well known fish. It is armed with spines like perch, but has only one back fin. It is of a dirty green, almost transparent, and spotted with black. It is found i shoals in the deep parts of running streams, and is esteemed good food. It seldom exceeds six inches in length.

THE HOLOCENTERS.

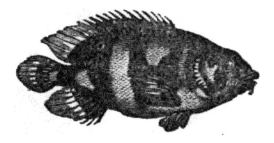

Nearly allied to the perch tribe are the Holocenters; one of the most remarkable of which genus is the Lanceola. ted Holocenter. It has a large head, with a mouth in

proportion; the bones of the lips are broad; the jaws are
of equal length, and armed with several rows of little sharp
teeth; as is the palate; but the tongue is smooth and moveable. The nostrils are double, the hinder pair near the
eyes. Hereabout begin the scales, which are small, tender,
and smooth. The pupil of the eye is black, the iris blue.
The front operculum is made of two small rounded plates,
of which the hinder one is strongly serrated. The gills
have a wide aperture, and one half of the membrane is concealed. The body is broad, the belly prominent, and the
anus in the middle of the body. The colour of the fish is
silvery with transverse stripes and spots of brown. The
soft rays of the fins are mostly divided into four branches.
This species is produced in the East Indies, and takes its
name from the shape of the fins. In the Indian ocean, and
on the coasts of Africa and America there are found several other varieties of the Holocenter, which are remarkable
for the brightness of their colours. The structure of their
mouths proves them to be carnivorous. They mostly prey
on crabs and young fish, which they swallow whole. Their
flesh is much esteemed by the natives, it being pleasant
and wholesome food:

THE STICKLE-BACK

Is a well known little fish. In the fens of Lincolnshire
they are found in such numbers, that they are used to manure the land. There are three species, the COMMON, or
THREE SPINED, the TEN SPINED, and the FIFTEEN SPINED.
The two first seldom reach the length of two inches; the
latter sometimes grows to that of six, and is found in the
sea only.

THE MACKEREL GENUS

Is distinguished by a number of small fins, between the
back fin and the tail. The common Mackerel is a beauti-

ful fish, which is well known for the seasonable visits which it pays to our shores. Nothing can equal the brilliancy of its colours, which are a fine green, varied with blue and black, and which death indeed impairs, but cannot totally destroy.

THE MACKEREL,

As well as the haddock and the whiting, are thought, by some, to be driven upon our coasts rather by their fears than their appetites; and it is to the pursuit of the larger fishes we owe their welcome visits. It is much more probable, that they come for that food which is found in more plenty near the sea shore than farther out at sea. The Mackerel emits a phosphoric light when fresh from the sea. When taken out of the water it soon dies, and even in the water, if it advance with too much impetuosity against the net. It is caught with that instrument, or with a hook baited with bits of red cloth, or small herrings, and pieces of other kinds of fish or flesh. In some places it is taken by lines from boats, as during a fresh gale of wind it readily seizes a bait: it is necessary that the boat should be in motion, in order to drag the bait along near the surface of the water. There is a great fishery for Mackerel on some parts of the west coast of England. This is of such an extent, as to employ in the whole a capital of nearly two hundred thousand pounds. The fishermen go out to the distance of several leagues from the shore, and stretch their nets, which are sometimes several miles in extent, across the tide, during the night. The meshes of these nets are just large enough to admit the heads of tolerable large fish, and catch them by the gills. A single boat has been known to bring in, after one night's fishing, a cargo that has sold for nearly seventy pounds.

THE TUNNY

RETAINS not only the character, but the habits of the mackerel. They resort in vast shoals to the Mediterranean at certain seasons, and, from the earliest periods of history, have constituted a considerable branch of commerce there. The Tunny, however, differs greatly from the mackerel in size. One which Mr. Pennant saw at Inverary in Scotland, weighed four hundred and sixty pounds. It was seven feet ten inches in length, and the circumference in the largest part was five feet seven, and near the tail only one foot six. The pieces, when fresh cut, appear like raw beef, but when boiled turn pale, and have something the flavour of salmon.

THE SCAD, OR HORSE MACKEREL,

Is much smaller than the tunny. It is distinguished by a large black spot on the covers of the gills, and by the second back fin reaching almost to the tail. It is tolerable food.

THE SURMULLET

HAS the body slender; the head almost four cornered; the fin covering the gills with three spines; some of these have beards; it was a fish highly prized by the Romans, and is still considered as a very great delicacy.

THE GURNARD GENUS

Is known by a slender body, the head nearly four cornered, and covered with a bony coat; the fin covering the gills

with seven spines: the pectoral and ventral fins strengthened with additional muscles and bones, and very large for
the animal's size.

Of the Gurnard Mr. Pennant has remarked five species.
The GRAY, the RED, the PIPER, the SAPPHIRINE, and the
STREAKED. They have all nearly the same nature and
manners. They are taken in deep water, with no other
bait than a red rag, and are esteemed good food.

THE BOW BANDED CHÆTODON.

THE head of this curious fish is large; the eyes are small
and placed near the top; the pupil is black, the iris gold
yellow. The aperture of the gills is wide, and at the covert there is a spine. The lateral line is made up of white
dots. The ground colour is brown, which towards the
back inclines to black; and looks as if covered with velvet,
and inlaid with ivory. The tail is not divided. This species inhabits the coasts of Brazil, and other parts of South
America, and grows from three to six inches in length.

In winter or the rainy seasons, they lie in deep holes
near the shore, which they quit in spring to come into the
shallows near the land: during the summer, when the sun
in those climates blazes the whole day with irresistible
fierceness, they keep at the depth of twenty or thirty yards,

which protects them from its intense heats. They spawn in the coldest time of the year: and being a lively fish, great numbers of the young fry are caught for the sake of being kept in vases, but in which they seldom come to maturity, and never increase.

The BEAKED CHÆTODON. This fish, which is by far the most curious of the tribe to which it belongs, frequents the shores and mouths of rivers in India, and about the Indian islands. Its length is somewhat more than six inches, and its colour is whitish, or very pale brown, with commonly four or five black bands running across the body, which is ovate and compressed. The dorsal and anal fins are very large, and on the former there is an eyelike spot of considerable magnitude. The snout is lengthened and cylindrical, and is the instrument by means of which the animal obtains its subsistence. Flies and other small insects that hover over the water, constitute the principal food of this fish. When it sees a fly on a plant, it slowly and cautiously approaches, as perpendicularly as possible under the object, puts its body in an oblique direction, with its mouth and eyes near the surface, fixes the latter on the insect, remains for a moment motionless, and then without showing its mouth above the surface, darts a drop of water from its tubular snout. So dexterously does it take its aim, that at the distance of four, five, or six feet, it hardly ever fails to bring the fly into the water. From this circumstance it derives the name which some naturalists give to it, of the Jaculator, or Shooting-fish.

THE PARROT FISH.

THE head of this fish is somewhat similar to that of the carp. The body is broad, the tail narrow. The ground colour is red, which is beautifully relieved by broad silvery stripes all along the body ; the belly is white. The fins are small; the scales broad, thin, finely radiated, and very loose. The pectoral, tail, and ventral fins are yellow at their origin, and gray at the extremities; and a kind of spine runs along the root of the ventral fin. This species is found in both Indies.

The fourth division of the spinous fishes consists of the ABDOMINAL, or those which have the *ventral fins behind the pectoral*, that is nearer the tail, as in the salmon.

THE LOACH

Is the first genus which is noticed in this division, and is a well known little fish, which never exceeds four inches in length. In its general aspect it has a pellucid appearance. It is distinguished by an oblong body; almost equally broad throughout; the head small, a little elongated: the eyes in the hinder part of the head; the fin covering the gills from four to six rays; the covers of the gills closed below. The back is mottled with irregular collections of small black dots.

THE FOUR-EYED LOACH.

THE head of this species is foreshortened and broader than it is high. The lower jaw is the longest, and it lengthens downwards, and not in front, like other fish. Both jaws, as well as the palate and tongue, are armed with teeth; the barbles arise from the corners or extremities of the upper lip. The nostrils are single, and near the mouth. The eyes are very remarkable: each containing two parts

or a double pupil, which has caused it to be called four eyes. The cavity of the eyes differs from other fish: this cavity is not a cylinder, as in other animals, but a part of one only; on each side at the top of the head there is an arched thin bone advancing towards the scull; these bones face each other with their concave surfaces; the eye is cylindrical, and is fixed in this cavity, but rises above it: the pupil is seen above the surface, enclosed in a black iris; as the cornea is equally luminous in the internal part, the pupil is seen double. The gill coverts are smooth and slippery; the body upwards is broader than it is thick; but towards the tail it takes a rounded form. The sides are ornamented with five longitudinal dark-brown stripes, which run quite to the tail, where the two outermost are connected by a transverse stripe, and the three middlemost by another. The lateral line is scarcely visible; the anus is nearer to the tail than to the head. The dorsal fin is small, and near the tail. All the fins, except the ventrals, are covered mostly with small scales; but on the body the scales are larger. It produces its young alive.

This singular fish is found in the rivers of Surinam, near the seacoast. It multiplies fast, and is esteemed by the inhabitants as good food; it grows from six to ten inches in length. Linnæus refers it to the loach genus, from which, however, it differs so much that Bloch makes a new genus of it, under the name of Anableps.

THE SALMON,

WHICH was known to the Romans, but not to the Greeks, is distinguished from other fish by having two dorsal fins, of which the hindermost is fleshy and without rays; they have teeth both in the jaws and the tongue, and the body is covered with round and minutely striated scales. Gray is the colour of the back and sides, sometimes spotted with black, and sometimes plain. The belly is silvery. It is entirely a northern fish, being found both at Greenland, Kamtschatka, and the northern parts of North America, but never so far south as the Mediterranean.* About the latter end of the year the Salmon begin to press up the rivers, even for hundreds of miles, to deposit their spawn, which lies buried in the sand till spring, if not disturbed by the floods, or devoured by other fishes. In this peregrination it is not to be stopped even by cataracts. About March

* Salmon are now scarce in all our rivers south of the Merrimack. In the Connecticut they were once so abundant as to be less esteemed than shad, and the fishermen used to require their purchasers to take some salmon with their shad. Within the memory of persons living, they were taken in plenty even as far up as Vermont. The Indians used to catch a great many of them, as they were ascending Bellows Falls. It is supposed that the locks, dams, and canals constructed in the river, have driven this valuable fish away.

the young ones begin to appear, and about the beginning of May the river is full of the Salmon fry, which are then four or five inches long, and gradually proceed to the sea. About the middle of June the earliest fry begin to return again from the sea, and are then from twelve to fourteen inches long. The growth of this fish is so extraordinary, that a young Salmon being taken at Warrington, and which weighed seven pounds on the 7th of February, being marked with a scissors on the back fin, was again taken on the 17th of March following, and was then found to weigh seventeen pounds and a half.

Rapid and stony rivers, where the water is free from mud, are the favourite places of most of the Salmon tribe, the whole of which is supposed to afford wholesome food to mankind. The chief English rivers in which Salmon are caught are the Thames, the Severn, the Trent, and the Tyne. The Scotch fisheries are very productive. These fish when taken out of their natural element very soon die ; to preserve their flavour they must be killed as soon as they are taken out of the water. The fishermen usually pierce them near the tail with a knife, when they soon die with loss of blood.

THE SEA-TROUT, OR SALMON-TROUT,

MIGRATES like the salmon up several of our rivers, spawns, and returns to the sea. The shape is thicker than the common Trout. The head and back are dusky, with a gloss of blue and green, and the sides, as far as the lateral line, are marked with large irregular spots of black. The flesh, when boiled, is red, and resembles that of the salmon in taste.

THE WHITE TROUT

APPEARS much of the same nature, and migrates out of the sea into the river Esk, in Cumberland, Eng. from July to September.

THE SAMLET

Is considered by Mr. Pennant as a distinct species, and not as the fry of the salmon, as some persons have sup-

posed. In this case it must be considered as the smallest of the trout genus, from which, however, it materially differs. It seldom exceeds six or seven inches in length.

THE TROUT.

It is a matter of surprise that this common fish has escaped the notice of all the ancients, except Ausonius. It is also singular, that so delicate a species should be neglected, at a time, when the folly of the table was at its height; and that the epicures should overlook a fish that is found in such quantities in the lakes of their neighbourhood, when they ransacked the universe for dainties. The milts of *murænæ* were brought from one place ; the livers of *scari* from another; and oysters even from so remote a spot as Sandwich : but there was and is a fashion in the article of good living. The Romans seem to have despised the Trout, the piper, and the doree ; and we believe Mr. Quin himself would have resigned the rich paps of a pregnant sow, the heels of camels, and the tongues of flamingos, though dressed by Heliogabalus's cooks, for a good jowl of salmon with lobster sauce.

The general shape of the Trout is rather long than broad; in several of the Scotch and Irish rivers, they grow so much thicker than in those of England, that a fish, from

eighteen to twenty-two inches, will often weigh from three to five pounds. This is a fish of prey, has a short round-ish head, blunt nose, and wide mouth, filled with teeth, not only in the jaws, but on the palate and tongue; the scales are small, the back ash colour, the sides yellow, and, when in season, it is sprinkled all over the body and cov-ers of the gills with small beautiful red and black spots: the tail is broad.

The colours of the Trout, and its spots, vary greatly in different waters, and in different seasons; yet each may be reduced to one species. In Llyndivi, a lake in South Wales, are Trouts called Cochy-dail; marked with red and black spots as big as sixpences; others unspotted, and of a reddish hue, that sometimes weigh near ten pounds, but are bad tasted.

In Lough Neagh, in Ireland, Trouts are called Buddaghs, which sometimes weigh thirty pounds.

Trouts (probably of the same species) are also taken in Ulleswater, a lake in Cumberland, of a much superior size to those of Lough Neagh. These are supposed to be the same with the Trout of the Lake of Geneva.

In the River Eyneon, not far from Machyntleth, in Me-rionethshire, Wales, and in one of the Snowdon lakes, are found a variety of Trout, which are naturally deformed, having a strange crookedness near the tail, resembling that of the perch before described.

The stomachs of the common Trouts are uncommonly thick and muscular. They feed on the shell-fish of lakes and rivers, as well as on the small fish. They likewise take into their stomachs gravel, or small stones, to assist in comminuting the testaceous parts of their food. The Trouts of certain lakes in Ireland, such as those of the province of Galway, and some others, are remarkable for the great thickness of their stomachs, which, from some slight resemblance to the organ of digestion in birds, have been called gizzards; the Irish name the species that has them GILLAROO TROUTS. These stomachs are sometimes served up to table under the former appellation. It does not, however, appear that the extraordinary strength of stomach in the Irish fish should give any suspicion that it is a distinct species; the nature of the waters might increase the thickness; or the superior quantity of shell-fish, which may more frequently call for the use of its comminuting powers than those of our Trouts, might occasion this difference.

Trouts are most voracious fish, and afford excellent diversion to the angler; the passion for the sport of angling is so great in the neighbourhood of London, that the liberty of fishing in some of the streams in the adjacent counties is purchased at the rate of ten pounds per annum.

These fish shift their quarters to spawn, and, like salmon, make up towards the head of rivers to deposit their roes.

THE RED CHAR.

THE head of this fish terminates in a blunt point, and its body is covered with very minute scales; the lateral line is straight. All the fins except the dorsal are reddish. This species is very properly denominated the Alpine Char, by Linnæus; for its constant residence is in the lakes of the high and mountainous parts of Europe. A few are found in some of the lakes in Wales, and in Loch Inch, in Scotland; from which last, it is said to migrate into the Spey to spawn. Seldom, however, does this spe-

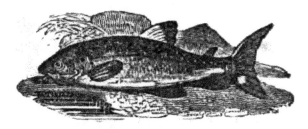

cies venture into any running stream; its principal resort is in the cold lakes of the Lapland Alps, where it is fed by the innumerable quantity of gnats that infest those dreary regions.

· The largest and most beautiful Chars are found in the Lake of Winander-Mere, in Westmoreland, Eng. where there are three species, the RED, the GILT, and the CASE CHAR. These kinds are nearly similar in their external appearance; but the time and manner of their spawning are very different. The method of taking these fish is with nets, or trammels as they are called, which are furnished with bait to allure the fish, and left for several days, till they are known to enter them. Potted Char is a delicacy which is in high repute on the Continent as well as in England.

THE GRAYLING, OR UMBER,

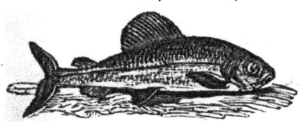

Is another of this genus, which haunts clear and rapid streams. It is of an elegant form, less deep than a trout.

It is in general of a fine silvery gray, but when just taken it is varied slightly with blue and gold. The scales are large; the first dorsal fin consists of twenty-one rays; this fin is spotted; all the rest are plain: the tail is much forked. It haunts clear and rapid streams, particularly those of mountainous countries. In Lapland, where it is very common, the inhabitants use its entrails, instead of rennet, to make their cheese from the milk of the rein-deer. The stomach is so hard and thick, that to the touch it appears like cartilage. The largest that has been heard of was taken near Ludlow; it was half a yard long, and weighed four pounds six ounces. The ancients believed that the oil from them would obliterate freckles and small pox marks.

THE SMELT

Inhabits the northern seas, and is never found so far south as the Mediterranean. Its name is supposed to be a contraction of "smell it," from its very agreeable smell. The Germans, however, call it the Stink-fish. Its form is very elegant; it is of a silvery colour, tinged with yellow; and the skin is almost transparent. The largest we have heard of was thirteen inches long, and weighed half a pound. There are two species, the Hepsetus and the Menidea.

THE GWINIAD

Is found in the lakes of several of the alpine parts of Europe. It is a gregarious fish, and approaches the shores in vast shoals in spring and summer. A fisherman, in 1775, took near eight thousand at one draught. It is of an insipid taste, and must be eaten soon. The back is arched and glossed with blue and purple, the sides are of a silvery cast, tinged with gold. The mouth is small, and without teeth. It is about eleven inches long.

THE PIKE

Is common in most of the lakes of Europe, but the largest are those taken in Lapland, which, according to Schæffer, are sometimes eight feet long. They are taken there in great abundance, dried and exported for sale. The largest fish of this kind which we have ever heard of in England, weighed thirty-five pounds.

According to the common saying, these fish were introduced into England in the reign of Henry the Eighth, in 1537. They were so rare, that a Pike was sold for double the price of a house-lamb, in February, and a Pickerel for more than a fat capon.

All writers who treat of this species bring instances of its vast voraciousness. We have known one that was choked by atempting to swallow one of its own species that proved too large a morsel. Yet its jaws are very loosely connected; and have on each side an additional bone like the jaw of a viper; which renders them capable of great distention when it swallows its prey. It does not confine itself to feed on fish and frogs; it will devour the water-rat, and draw down the young ducks as they are swimming about.

At the Marquis of Stafford's canal at Trentham, Eng. a Pike seized the head of a swan, as she was feeding under water, and gorged so much of it as killed them both. The servants perceiving the swan with its head under water for a longer time than usual, took the boat, and found both swan and Pike dead.

But there are instances of its fierceness still more surprising, and which, indeed, border a little on the marvellous. Gesner relates, that a famished Pike in the Rhone seized on the lips of a mule, that was brought to water, and that the beast drew the fish out before it could disengage itself: that people have been bit by these voracious creatures while they were washing their legs; and that they will even contend with the otter for its prey, and endeavour to force it out of its mouth.

Small fish show the same uneasiness and detestation at the presence of this tyrant, as the little birds do at the sight of the hawk or owl. When the Pike lies dormant near the surface (as is frequently the case) the lesser fish are often observed to swim around it in vast numbers, and in great anxiety. Pike are often haltered in a noose, and taken while they lie thus asleep, as they are often found in the ditches near the Thames, in the month of May.

In the shallow water of the Lincolnshire fens, in England, they are frequently taken in a manner peculiar, we believe, to that county and the isle of Ceylon. The fishermen make use of what is called a crown-net, which is no more than a hemispherical basket, open at top and bottom. He stands at the end of one of the little fen-boats, and frequently puts his basket down to the bottom of the water, then poking a stick into it, discovers whether he has any booty by the striking of the fish: vast numbers of Pike are taken in this manner.

The longevity of this fish is very remarkable, if we may credit the accounts given of it. Rzaczynski tells us of one that was ninety years old; but Gesner relates, that in the year 1497, a Pike was taken near Hailbrun, in Suabia, with a brazen ring affixed to it, on which were these words in Greek characters: *I am the fish which was first of all put into this lake by the hands of the governor of the universe, Frederick the Second, the 5th of October,* 1230: so that the former must have been an infant to this Methuselah of a fish.

Pike spawn in March or April, according to the coldness or warmth of the weather. When they are in high season, their colours are very fine, being green, spotted with bright yellow; and the gills are of a most vivid and full red. When out of season, the green changes to gray, and the yellow spots turn pale.

The head is very flat; the upper jaw broad, and is shorter than the lower: the under jaw turns up a little at the end, and is marked with minute punctures. The teeth are very sharp, disposed not only in the front of the upper jaw, but in both sides of the lower, in the roof of the mouth, and often the tongue. The slit of the mouth, or the gape, is wide; the eyes small. The dorsal fin is placed very low on the back, and consists of twenty-one rays; the pectoral of fifteen; the ventral of eleven; the anal of eighteen. The tail is bifurcated.

THE GAR-PIKE, GAR-FISH, OR SEA NEEDLE,

COMES in shoals on our coasts, and precedes the mackerel, whose shoals it is vulgarly supposed to pilot through the regions of the deep. It resembles that fish in flavour, but is distinguished from all of the kind by the back bone, which turns a fine light green when the fish is boiled. It sometimes grows to the length of three feet. The jaws are exceeding long, slender, and pointed, and the edges of them are armed with numbers of short slender teeth. It is sometimes known by the name of the Horn-fish. The tail is forked. The back is a fine green, beneath which appears a rich changeable purple and blue, and the sides and belly are of a fine silvery hue.

THE SAURY-PIKE

Is about eleven inches long, and its jaws are protracted like those of the sea-needle. The body also resembles that of an eel, but, like the mackerel, it has a number of small fins near the tail, which is forked.

THE ARGENTINE

Is a small fish between two and three inches long. The body is compressed, and almost of an equal breadth to the anal fin. The back is of a dusky green, the sides and covers of the gills as if plated with silver. It is taken in the sea.

THE ATHERINE

Is common in the sea near Southampton, England, where it is called a smelt. It never deserts the place, and is constantly taken, except in hard frost. It is about four inches in length, the back straight, the belly a little protuberant. Its colour is silvery, tinged with yellow, and below the side line is a row of black spots. It is semi-pellucid.

THE MULLET

Was formerly much celebrated as a treat for the epicure, and frequent allusions to it are found in the ancient satirists. It is a fish of an elegant form; is generally found by the sea shores, where it roots like a hog in the sand or mud, and it is so active, that it frequently escapes, by leaping out of the fishermen's nets. The head is almost square, and is flat at the top. It has no teeth, only in the upper lip is a small roughness. The tail is much forked. The colour of the back is dusky, marked with blue and green. The sides silvery, marked with dusky lines, reaching from the head to the tail. The belly is silvery.

Of the FLYING FISH so much has been said under the article Dorado, that it is only necessary in this place to add a short description of it.

The body of this fish is oblong; the head is almost three cornered; the fin covering the gills with ten rays; the pectoral fins placed high, and as long as the whole body; the back fin at the extremity of the back. The tail is bifurcated.

THE HERRING.

The common Herring is distinguished from the other fish of the same tribe, by the projection of the lower jaw, which is curved, and by having seventeen rays in the ventral fin. The head and mouth are small, the tongue short, pointed, and armed with teeth; the covers of the gills generally have a violet or red spot, that disappears soon after the death of the fish, which survives a very short time, when taken out of its natural element.

The principal of the British Herring fisheries are off the Scotch and Norfolk coasts; the fishing is carried on by nets stretched in the water, one side of which is kept from sinking, by means of buoys fixed to them at proper distances; and, as the weight of the net makes the side sink to which no buoys are fixed, it is suffered to hang in a perpendicular position, like a screen; and the fish, when they endeavour to pass through it, are entangled in its meshes, from which they cannot disengage themselves. There they remain till the net is hauled in, and they are shaken or picked out. The nets are never stretched to catch Herrings but during the night, for in the dark they are to be taken in much the greatest abundance.

After the nets are hauled, the fish are thrown upon the deck of the vessel, and each of the crew has a certain task assigned to him. One part is employed in opening and gutting them, another in salting, and a third in packing them in the barrels in layers of salt. The Red Herrings lie twenty-four hours in the brine; they are then taken out, strung by the gills on little wooden spits, and hung in a chimney formed to receive them; after which a fire of brushwood, which yields much smoke, but no flame, is kindled under them, and they remain there till sufficiently smoked and dried, when they are put into barrels for carriage.

Herrings become very soon tainted after they are dead; in summer they are sensibly worse for being out of the water only a few hours; and if exposed only a few minutes to the rays of the sun, they are quite useless, and will not take the salt.

The PILCHARD is thicker and rounder than the Herring.

The nose is shorter in proportion, and turns up. The back is more elevated, the belly less sharp. The back fin of the Pilchard is placed exactly in the centre of gravity, so that when taken up by it, the fish exactly preserves an equilibrium, whereas that of the Herring dips at the head. The scales of the Pilchard adhere very closely, whereas those of the Herring very easily drop off. The Pilchard is in general less than the Herring, and is fatter and fuller of oil.

Of all the migrating fish, the Herring and the Pilchard take the most adventurous voyages. Herrings are found in the greatest abundance in the highest northern latitudes. In those inaccessible seas, that are covered with ice for a great part of the year, the Herring and Pilchard find a quiet and sure retreat from all their numerous enemies: thither neither man, nor their still more destructive enemy, the fin-fish, or the cachalot, dares to pursue them. The quantity of insect food which those seas supply is very great; whence, in that remote situation, defended by the icy rigour of the climate, they live at ease, and multiply beyond expression. From this most desirable retreat, Anderson supposes they would never depart, but that their numbers render it necessary for them to migrate ; and, as bees from a hive, they are compelled to seek for other retreats.

For this reason, the great colony is seen to set out from the icy sea about the middle of winter; composed of such numbers, that if all the men in the world were to be loaded with Herrings, they would not carry the thousandth part away. But they no sooner leave their retreats, but millions of enemies appear to thin their squadrons. The finfish and the cachalot swallow barrels at a yawn; the porpus, the grampus, the shark, and the whole numerous tribe of dog-fish, find them an easy prey, and desist from making

H3

war upon each other: but still more, the unnumbered
flocks of sea-fowl that chiefly inhabit near the pole, watch
the outset of their dangerous migration, and spread extensive ruin.

In this exigence, the defenceless emigrants find no other
safety, but by crowding closer together, and leaving to
the outmost bands the danger of being the first devoured;
thus like sheep when frightened, that always run together in a body, and each finding some protection in being
but one of many that are equally liable to invasion, they
are seen to separate into shoals, one body of which moves
to the west, and pours down along the coast of America,
as far south as Carolina, and but seldom farther.   In Chesapeake Bay, the annual inundation of these fish is so great,
that they cover the shores in such quantities as to become
a nuisance.   Those that hold more to the east, and come
down towards Europe, endeavour to save themselves from
their merciless pursuers, by approaching the first shore
they can find; and that which first offers in their descent
is the coast of Iceland, in the beginning of March.   Upon
their arrival on that coast, their phalanx, which has already
suffered considerable diminutions, is, nevertheless, of amazing extent, depth, and closeness, covering an extent of
shore as large as the island itself.   The whole water seems
alive; and is seen so black with them at a great distance,
that the number seems inexhaustible.

That body which comes upon the English coast, begins
to appear off the Shetland Isles in April.   These are the
forerunners of the grand shoal which descends in June;
while its arrival is easily announced, by the number of its
greedy attendants, the gannet, the gull, the shark, and the
porpus.   When the main body is arrived, its breadth and
depth is such, as to alter the very appearance of the ocean.
It is divided into distinct columns, of five or six miles in

length, and three or four broad; while the water before
them curls up, as if forced out of its bed. Sometimes
they sink for the space of ten or fifteen minutes, then rise
again to the surface : and, in bright weather, reflect a variety of splendid colours, like a field bespangled with purple, gold, and azure. The fishermen are ready prepared to
give them a proper reception; and, by nets made for the
occasion, they take sometimes above two thousand barrels
at a single draught.

Such has long been the received opinion with respect
to the migration of Herrings, and it is so poetical that it is
almost a pity to disturb it. But science must listen only
to the voice of truth. The author of the British Naturalist has given a rude shock to the migratory theory. " Simply, then, the story cannot be true (says he), because it is
impossible. The Herrings do not come in myriads from
the polar sea, beginning their progress in January, because
there are no means of producing them there. Spawn has
not been found to animate in any place except floating
near the surface, or in shallow water, where both the sun
and the air act upon it; and while the polar seas and shores
are open to such action, the Herrings are not there; they
are on our shores, the full grown and the young. But setting aside the impossibility, the supposed emigration would
be without an object: they would not come for food, as
they are said to leave the north just when food would be
found there; and if they are annually produced in the north,
they could not come to our shores for the purpose of spawning, even though they are all obviously in preparation for
such a purpose. Beside, there is no animal that migrates
southward in the spring; and therefore the theory would
require one law for the rest of creation, and another for
the Herring; that the latter should be chilled by the genial warmth of the spring, and warmed by the polar frost.

Now, so far is the production of fish from being independent of the influence of heat, that, just as one would be led to infer from the slow progress of the solar beams through the element in which they live, they require the whole, or the greater part of our summer, to mature the germs of their countless broods. Nay, it appears that many, if not most of the species, cannot mature their spawn in the depths of the ocean, to which they retire to recruit their strength, but that they come to the shores and shallows, where the heat of the sun can penetrate to the bottom, and be reflected by it, for the purpose of maturing as well as depositing their spawn.

" The Herrings come to the shores and estuaries to mature and propagate their spawn, which they do over a greater range of the year than most other fish ; continuing the operation to the middle of winter, and retiring into deeper water after that is done. But there is no reason to conclude, that they have much migration in latitude ; or, that they ever move far from those shores which they frequent in the season. The fry too are found on the shores and in the bays and estuaries frequented by their parents ; and they do not go to the deep water till late in the season. They even appear to go farther up the rivers than the old fish, for they may be taken in brackish water, with a common trout fly."

THE SPRAT

Is now generally allowed not to be the fry of the herring, as, from its great resemblance, was formerly supposed.

The back fin of the Sprat is more remote from the nose than that of the herring; but a principal distinction is, that the belly of both the herring and pilchard is quite smooth, whereas that of the Sprat is serrated. The herring has fiftysix vertebræ, the Sprat only fortyeight. The Sprat seldom exceeds five inches in length.

THE ANCHOVY

Is about six inches and a half in length. The body is slender, but thicker in proportion than the herring. The scales are large, and easily fall off. The back is green, and semipellucid; the sides and belly silvery; and the tail forked.

At different seasons it frequents the Atlantic Ocean and the Mediterranean Sea, passing through the Straits of Gibraltar towards the Levant in the months of May, June, and July. The greatest fishery is at Gorgona, a small isle west of Leghorn, where they are taken at night in nets, into which they are allured by lights fixed to the stern of the vessels. When cured, their heads are cut off, their gall and entrails taken out, then salted and packed in barrels. It scarce needs to be mentioned that, being put on the fire, they dissolve in almost any liquor. They are well tasted when fresh. But it has been found by experience, that Anchovies taken thus by torch light are neither so good, so firm, nor so proper for keeping, as those which are taken otherwise. From December to March, vast numbers are caught on the shores of Provence and Catalonia, and during June and July in the English Channel, and in the environs of Bayonne, Venice, Rome, and Genoa.

From the Anchovy the ancients prepared one of the liquids called *garum*, which was in high repute among epicures.

THE SHAD

Is taken in many rivers; those of the Severn are most esteemed in England, and are distinguished by the London fishmongers by the French name of *Alosse*. The Thames Shad is a very insipid coarse fish. The Thames Shad, when it visits the Severn, is called the *Twaite*, and is held in great disrepute.

The difference between the two kinds is as follows:— The true *Shad* weighs from four to eight pounds; the *Twaite* from half a pound to two. The Twaite may also be known from a small Shad, by having one or more black spots on the sides: when it has only one, it is always near the gill.

The Shad of America is a very superior fish, and is abundant in all the northern rivers. Those of the Connecticut are particularly esteemed, and when salted and barrelled, command a high price. These fish are chiefly taken during the months of April and May. They ascend the rivers for many miles, and formerly large numbers of them were caught in the Connecticut, at the distance of 200 miles from its mouth.

The Shad in form rather resembles the herring, but is larger and thinner, or more compressed in proportion. The head slopes considerably from the back; and the under jaw is longer than the upper.

THE CARP

Is a genus which, besides the fish which bears that name, includes several others well known to anglers, viz. the barbel, the gudgeon, the crucian, the bream, the tench, the roach, dace, &c.

' THE CARP

Is one of the naturalized fish in England, having been introduced there by Leonard Maschal, about the year 1514, to whom the English were also indebted for that excellent apple the pepin. The many good things which that island wanted before that period are enumerated in this old distich:

> "Turkeys, carp, hops, pickerel, and beer,
> Came into England all in one year."

As to the two last articles we have some doubts, the others we believe to be true. Russia wants these fish at this day; Sweden has them only in the ponds of the people of fashion: Polish Prussia is the chief seat of the Carp; they abound in the rivers and lakes of that country, particularly in the Frisch and Curischhaff, where they are taken of a vast size. They are there a great article of commerce, and sent in well boats to Sweden and Russia. The merchants purchase them out of the waters of the noblesse of the country, who draw a good revenue from this article. Neither are there wanting among our gentry instances of some who make good profits of their ponds.

The ancients do not separate the Carp from the sea-fish. We are credibly informed that they are sometimes found in the harbour of Dantzick, between the town and a small place called Hela.

Carp are very long-lived. Gesner brings an instance of one that was a hundred years old. They also grow to a very great size. On our own knowledge we can speak of

none that exceed twenty pounds in weight; but Jovius
says, that they were sometimes taken in the Lacus Larius
(the Lago di Como) of two hundred pounds weight; and
Rzaczynski mentions others taken in the Dneister that
were five feet in length.

The Carp is a prodigious breeder: its quantity of roe
has been found so great that, when taken out and weighed
against the fish itself, the former has been found to pre-
ponderate. From the spawn of this fish caviare is made
for the Jews, who hold the sturgeon in abhorrence.

These fish are extremely cunning, and on that account
are by some styled the *River Fox*. They will sometimes
leap over the nets, and escape that way; at others, will
immerse themselves so deep in the mud, as to let the net
pass over them. They are also very shy of taking a bait;
yet at the spawning time they are so simple as to suffer
themselves to be tickled, and caught by any body that will
attempt it. It is so tenacious of life that it may be kept
alive for a fortnight in wet straw or moss.

This fish is apt to mix its milt with the roe of other fish,
from which is produced a spurious breed; we have seen
the offspring of the Carp and tench, which bore the greatest resemblance to the first: we have also heard of the
same mixture between the Carp and bream.

The Carp is of a thick shape; the scales very large, and
when in best season of a fine gilded hue. The jaws are
of equal length; there are two teeth in the jaws, or on
the tongue; but at the entrance of the gullet, above and
below, are certain bones that act on each other, and com-
minute the food before it passes down. On each side of
the mouth is a single beard; above those on each side
another, but shorter: the dorsal fin extends far towards the
tail, which is a little bifurcated; the third ray of the dorsal fin is very strong, and armed with sharp teeth, pointing

downwards; the third ray of the anal fin is constructed in the same manner.

THE BARBEL

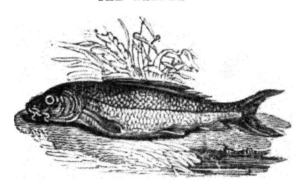

Was so extremely coarse as to be overlooked by the ancients till the time of Ausonius, and what he says is no panegyric on it; for he lets us know it loves deep waters, and, that when it grows old it was not absolutely bad.

It frequents the still and deep parts of rivers, and lives in society, rooting like swine with their noses in the soft banks. It is so tame as to suffer itself to be taken with the hand; and people have been known to take numbers by diving for them. In summer they move about during night in search of food, but towards autumn, and during winter, confine themselves to the deepest holes.

They are the worst and coarsest of fresh-water fish, and seldom eaten but by the poorer sort of people, who sometimes boil them with a bit of bacon, to give them a relish. The roe is very noxious, affecting those who unwarily eat of it with a nausea, vomiting, purging, and a slight swelling.

It is sometimes found of the length of three feet, and eighteen pounds in weight: it is of a long and rounded form; the scales not large. Its head is smooth; the nostrils placed near the eyes; the mouth is placed below: on

each corner is a single beard, and another on each side the
nose. The dorsal fin is armed with a remarkably strong
spine, sharply serrated, with which it can inflict a very severe wound on the incautious handler, and even do much
damage to the nets. The pectoral fins are of a pale brown
colour; the ventral and anal tipped with yellow; the tail
a little bifurcated, and of a deep purple; the side line is
straight. The scales are of a pale gold colour, edged with
black; the belly is white.

THE TENCH

UNDERWENT the same fate with the barbel, in respect to
the notice taken of it by the early writers: and even Ausonius, who first mentions it, treats it with such disrespect
as evinces the great capriciousness of taste; for that fish,
which at present is held in such good repute, was in his
days the repast only of the canaille.

It has been by some called the Physician of the fish, and
the slime so healing, that the wounded apply it as a styptic.

Whatever virtue its slime may have to the inhabitants
of the water, we will not vouch for; but its flesh is a wholesome and delicious food to those of the earth. The Germans are of a different opinion. By way of contempt they
call it Shoemaker. Gesner even says, that it is insipid
and unwholesome.

It does not commonly exceed four or five pounds in weight, but we have heard of one that weighed ten pounds. Salvianus speaks of some that arrived at twenty pounds.

They love still waters, and are rarely found in rivers; they are very foolish, and easily caught.

The Tench is thick and broad in proportion to its length; the scales are very small, and covered with slime. The irides are red; there is sometimes, but not always, a small beard at each corner of the mouth. The colour of the back is dusky; the dorsal and ventral fins of the same colour; the head, sides, and belly of a greenish cast, most beautifully mixed with gold, which is in its greatest splendour when the fish is in the highest season. The tail is quite even at the end, and very broad.

THE GUDGEON.

ARISTOTLE mentions the Gudgeon in two places; once as a river fish, again as a species that was gregarious; and in a third place he describes it as a sea fish.

This fish is generally found in gentle streams, and is of a small size; those few, however, that are caught in the Kennet and Coln rivers in England, are three times the weight of those taken elsewhere. The largest we ever heard of was taken near Uxbridge, Eng. and weighed half a pound.

They bite eagerly, and are assembled by raking the bed of the river; to this spot they immediately crowd in shoals, expecting food from this disturbance.

The shape of the body is thick and round; the irides tinged with red, the gill covers with green and silver. The lower jaw is shorter than the upper; at each corner of the mouth is a single beard; the back olive, spotted with black; the side line straight; the sides beneath that silvery; the belly white. The tail is forked; that, as well as the dorsal fin, is spotted with black.

THE BREAM

Is an article of great importance on the Continent, in Europe, though its flesh is not equal to that of the carp. It is found in all the great lakes, and in rivers which have a gentle current, and a bottom composed of marl, clay, and herbage; and it abides in the deepest parts. It is taken mostly under the ice; and this fishery is so considerable that, in some of the lakes belonging to Prussia, there have been taken to the value of two hundred pounds at a time; they are also caught in great quantities in Holstein, Mecklenburg, Livonia, and Sweden: in a lake near Nordkiœping, there were taken at one time in March, 1749, no less than fifty thousand, weighing eighteen thousand two hundred pounds.

It is extremely deep, and thin in proportion to its length. The back rises much, and is very sharp at the top. The head and mouth are small; on some we examined in the spring were abundance of minute whitish tubercles, an accident which Pliny seems to have observed befalls the fish of the Lago Maggiore and Lago di Como. The scales are very large; the sides flat and thin. The dorsal fin has eleven rays, the second of which is the longest: that fin, as well as all the rest, are of a dusky colour; the back of the same hue; the sides yellowish. The tail is very large, and of the form of a crescent.

THE RUD

Is found in the Charwell, near Oxford, Eng. in the Fens, near Holderness. The body is extremely deep, like that of the bream, but much thicker. The head is small; the back vastly arched; the scales very large. The back is of an olive colour; the sides and belly gold; the ventral and anal fins, and the tail, of a deep red. It appears to be the same fish with the Shallow of the Cam.

THE CRUCIAN

Is common in many of the fish-ponds about London, and other parts of the south of England; but we believe is not a native fish.

It is very deep and thick; the back is much arched; the dorsal fin consists of nineteen rays; the two first strong and serrated: the pectoral fins have each thirteen rays; the ventral nine; the anal seven or eight: the lateral line is parallel with the belly; the tail almost even at the end. The colour of the fish in general is a deep yellow; the meat is coarse, and little esteemed.

THE ROACH.

' SOUND as a roach,' is a proverb that appears to be but indifferently founded, that fish being not more distinguished

for its vivacity than many others; yet it is used by the
French as well as the English, who compare people of
strong health to the *Rouget*, or Roach.   It is so silly a fish
that it is called the Water Sheep.

It is a common fish, found in many of our deep still rivers, affecting, like the others of this genus, quiet waters.
It is gregarious, keeping in large shoals.   We have never
seen them very large.   Old Walton speaks of some that
weighed two pounds.   In a list of fish sold in the London
markets, with the greatest weight of each, communicated
to the Editor by an intelligent fishmonger, there is mention
of one whose weight was five pounds.

The Roach is deep, but thin, and the back is much elevated, and sharply ridged; the scales large, and fall off
very easily.   The side line bends much in the middle towards the belly.

THE DACE,

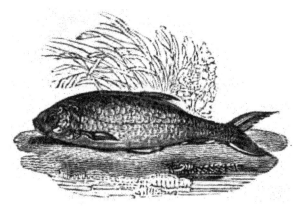

LIKE the roach, is gregarious, haunts the same places, is a

great breeder, very lively, and during summer is very fond of frolicking near the surface of the water. This fish, and the roach, are coarse and insipid meat.

Its head is small; the irides of a pale yellow; the body long and slender; its length seldom above ten inches, though in the above-mentioned list is an account of one that weighed a pound and a half; the scales smaller than those of the roach. The back is varied with dusky, with a cast of a yellowish green; the sides and belly silvery; the dorsal fin dusky; the ventral, anal, and caudal fins red, but less than those of the former; the tail is very much forked.

THE CHUB.

SALVIANUS imagines this fish to have been the *Squalus* of the ancients, and grounds his opinion on a supposed error in a certain passage in Columella and Varro, where he would substitute the word *Squalus* instead of *Scarus.*

That the *Scarus* was not our *Chub* is very evident; not only because the Chub is entirely an inhabitant of fresh waters, but likewise it seems improbable that the Romans would give themselves any trouble about the worst of river fish, when they neglected the most delicious kinds: all their attention was directed towards those of the sea; the difficulty of procuring them seems to have been the criterion of their value, as is ever the case with effete luxury.

The Chub is a very coarse fish, and full of bones; it frequents the deep holes of rivers, and during summer com-

monly lies on the surface, beneath the shade of some tree
or bush.　It is a timid fish, sinking to the bottom on the
least alarm, even at the passing of a shadow, but they
will soon resume their situation.　It feeds on worms, cat-
erpillars, grasshoppers, beetles, and other coleopterous insects that happen to fall into the water; and it will even
feed on cray fish.　This fish will rise to a fly.

This fish takes its name from its head, not only in the
English, but in other languages; it is called *Chub*, accord-
ing to Skinner, from the old English, *cop*, a head; the
French, in the same manner, call it *Testard*; the Italians,
Capitone.

It does not grow to a large size : we have known some
that weighed above five pounds; but Salvianus speaks of
others that were eight or nine pounds in weight.

The body is oblong, rather round, and of a pretty equal
thickness the greatest part of the way; the scales are
large: the irides silvery; the cheeks of the same colour;
the head and back of a deep dusky green; the sides silvery, but in the summer yellow; the belly white; the pectoral fins of a pale yellow; the ventral and anal fins red;
the tail a little forked, of a brownish hue, but tinged with
blue at the end.

THE BLEAK

Is very common in many of our rivers, and they keep together in large shoals.　These fish seem at certain seasons
to be in great agonies; they tumble about near the surface of the water, and are incapable of swimming far from
the place, but in about two hours recover, and disappear.
Fish thus affected, the Thames fishermen call Mad Bleaks.
They seem to be troubled with a species of *gordius* or
hair-worm, of the same kind with those which Aristotle
says that the *ballerus* and *tillo* are infested with, which

torments them so that they rise to the surface of the water, and then die.

Artificial pearls are made with the scales of this fish, and we think of the dace. They are beaten into a fine powder, then diluted with water, and introduced into a thin glass bubble, which is afterwards filled with wax. The French were the inventors of this art. Dr. Lister says, that when he was at Paris, a certain artist used in one winter thirty hampers full of fish in this manufacture.

The Bleak seldom exceeds five or six inches in length; their body is slender, greatly compressed sideways, not unlike that of the sprat. The eyes are large; the irides of a pale yellow; the under jaw the largest; the lateral line crooked; the gills silvery; the back green; the sides and belly silvery; the fins pellucid; the scales fall off very easily; the tail much forked.

During the month of July there appear in the Thames, in England near Blackwall and Greenwich, innumerable multitudes of small fish, which are known to the Londoners by the name of WHITE BAIT. They are esteemed very delicious when fried with fine flour, and occasion, during the season, a vast resort of the lower order of epicures to the taverns contiguous to the places they are taken at.

There are various conjectures about this species, but all terminate in a supposition, that they are the fry of some fish, but few agree to which kind they owe their origin. Some attribute it to the shad, others to the sprat, the smelt, and the Bleak. That they neither belong to the shad, nor the sprat, is evident from the number of branchiostegous rays, which in those are eight, in this only three. That they are not the young of the smelt is as clear, because they want the *pinna adiposa*, or rayless fin; and that they are not the offspring of the Bleak is extremely probable since we never heard of the White Bait being found in any other river, notwithstanding the Bleak is very common in several of the British streams: but as the White Bait bears a greater similarity to this fish than to any other we have mentioned, we give it a place here as an appendage to the Bleak, rather than form a distinct article of a fish which it is impossible to class with certainty.

It is probable that it is of the carp or *cyprinus* genus; it has only three branchiostegous rays, and only one dorsal fin; and in respect to the form of the body, it is compressed like that of the Bleak. Its usual length is two inches; the under jaw is the longest; the irides silvery, the pupil black; the dorsal fin is placed nearer to the head than to the tail, and consist of about fourteen rays; the side line is straight; the tail is forked; the tips black. The head, sides, and belly are silvery; the back tinged with green.

THE MINNOW

Is frequently found in small gravelly streams, where they keep in shoals.

The body is slender and smooth, the scales being extremely small. It seldom exceeds three inches in length. The lateral line is of a golden colour; the back flat, and of a deep olive; the sides and belly vary greatly in different fish; in a few are of a rich crimson, in others bluish, in others

white. The tail is forked, and marked near the base with a dusky spot.

THE GOLD FISH.

THESE fish are now quite naturalized in Europe, and breed as freely in the open waters as the common carp.

They were first introduced into England about the year 1691, but were not generally known till 1728, when a great number were brought over, and presented first to Sir Matthew Dekker, and by him circulated round the neighbourhood of London, whence they have been distributed to most parts of the country.

In China the most beautiful kinds are taken in a small lake in the province of Che-Kyang. Every person of fashion keeps them for amusement, either in porcelain vessels, or in the small basons that decorate the courts of the Chinese houses. The beauty of their colours, and their lively motions, give great entertainment, especially to the ladies, whose pleasures, by reason of the cruel policy of that country, are extremely limited.

In the form of the body they bear a great resemblance to a carp. They have been known in Europe to arrive at the length of eight inches; in their native place they are said to grow to the size of our largest herring.

The nostrils are tubular, and form a sort of appendages above the nose; the dorsal fin and the tail vary greatly in shape; the tail is naturally bifid, but in many is trifid, and in some even quadrifid; the anal fins are the strongest characters of this species, being placed not behind one another, like those of other fish, but opposite each other, like the ventral fins.

The colours vary greatly; some are marked with a fine blue, with brown, with bright silver; but the general predominant colour is gold, of a most amazing splendor; but their colours and form need not be dwelt on, since those who want opportunity of seeing the living fish, may survey them expressed in the most animated manner, in the works of Mr. George Edwards.

The SILVER FISH is a native of the seas in the vicinity of the Cape of Good Hope, and is about the size and shape of a small carp, which it also resembles in taste. It is of a white colour, transversely striped with silvery lines.

THE TELESCOPE CARP.

The whole body of this fish, and the ground colour of the fins, are of a beautiful sanguineous red, darker towards

the back, and lighter towards the belly; the membranes
of the fins are almost white, and the red rays shining
through them have a very fine effect; the three white
points of the tail form to the idea a trident or a tulip.
The head is short but large, the mouth is small, the nostrils are single. The pupil of the eye is black, the iris
yellow; and the eyes are protuberant; the back is round;
the lateral line nearer the back than the head. The scales
are large; the rays of the fins ramified. This beautiful
fish is found in the fresh waters of China, where it is kept
like the gold fish, of which it is probably a variety.

THE HORNED SILURE.

THIS fish has a broad, flat, thin head; and the horns,
which occupy the place of eyes in other species, are armed with short crooked spines like teeth, and are probably
weapons of defence. The head and body are entirely
covered with a skin resembling leather; the eyes lie on
each side of the head near the mouth; and the tail fin is
slightly forked.

This fish is of a very dark green colour; but the sides
are somewhat lighter. Another remarkable peculiarity
in this fish is the dorsal fin: it is close to the head, and
its front ray is long, stiff, dentated like the horns, and is
probably an instrument of defence also. The belly is
short and thick; and the lateral line goes meandering
along the middle of the body, and puts out branches each

way. This species grows to a considerable size; its flesh is eatable, but not much sought after. It is found near the shores of Asia and Surinam.

One of these fish caught at Surinam, on examination, was observed to have its mouth filled with yellow eggs, in none of which, however, could be found a fish completely formed; from which it is concluded, that the Silure, to defend her eggs from the voracious tribes, hatches them instinctively in her mouth. But she is supposed at times to emit them from her mouth, when in search of food to appease her appetite, and when satisfied, to take them into her mouth again.

Of the Silure tribe, there are nearly thirty different species, most of them natives of the Indian and American seas. One of them, the SILURIS CLARIAS, of Linnæus, is called Scheilan by the Arabians. Hasselquist relates, that he saw the cook of a Swedish merchant ship die of the poison communicated by a prick from the bone of the breast fin of this fish.

THE TRUMPET FISH.

THIS fish, which is also called the Bellows Fish, belongs to the genus Centriscus, and has a short broad body, laterally compressed, much resembling a pair of bellows, and of a pale red colour. The head, broadish above, ends in a bent cylinder below, and the aperture of the mouth, which is small, is at the end of the long beak; the aperture is closed by the lower jaw, which shuts into the upper like the lid of a snuff-box. The nostrils are double, and near the eyes, which are large, with a black pupil and a pale red iris. This fish can hide its small ventral fins in a furrow which lies behind them: all the fins are gray. The body is scaly and rough. This fish haunts the Mediterranean; but is often found in the ocean, whither, however, it seems to be driven by tempests, as it is never seen there but in stormy weather. It is tender, well tasted, and easy of digestion; but, being very thin, it is generally sold with other small fish at a low price. As the fins are very small in proportion to the other parts, so that it cannot swim fast enough to avoid its enemies, Providence has given it a moveable serrated spine for its defence, which is the first ray in the dorsal fin; with this it will successfully defend itself against almost any fish, unless taken by surprise. The name of the Trumpet-fish was given to it by mariners, in consequence of its breathing the water out of its snout with a sounding noise.

THE TOBACCO-PIPE FISH.

THIS species is known in the seas of America and Japan. The head is very long, quadrangular, and adorned with rays. The aperture of the mouth is wide, and in an oblique direction; the lower jaw is somewhat longer than the upper; the teeth are small; the tongue moveable; the nostrils double, and near the eyes, which are large, with a black pupil and silvery iris. The body is devoid of visible scales, flat in the front part, and rounded towards the tail. The lateral line has a straight direction. The anus is much nearer the tail-fin than the head; the belly is long; the fins are short and of a pale red, the rays mostly four-branched. This fish is brown, spotted with blue on the back and sides, and the belly is silvery.

Its usual length appears to be from twelve to eighteen

12

inches, exclusive of the bristle-formed process; which is usually one fourth as long as the body, but it is said to grow sometimes to between three and four feet. In some instances it has been found with two filiform appendages. It lives on small fry and lobsters; it is very plentiful, but lean, and is therefore only eaten by the poor.

THE SEA-COCK.

THIS fish has a very thin body, which is of a silver colour, inclining to red, and without visible scales. The head is large, very much shelving; the mouth large; the jaws are furnished with very small teeth, and the upper lip with two large bones. The nostrils are double, and near the eyes, which are round and large; the pupil is black, the iris brown, inclining to a silver gray. The opening of the gills is large; the covert is long, consisting of one plate, under which the membrane is hid. The lateral line is crooked at its origin; the anus is not far from the ventral fins. All the fins are of a bright green colour; in the dorsal fin, the nine first rays are short and hard, the next four long and soft, and both are single; the pectoral, ventral, and tail fins are branched.

This fish lives in all climates; it being found, according

to different authors, at Brazil, Jamaica, the Antilles, the East Indies, and Malta. It grows from six to nine inches in length, is well tasted, and lives upon worms, insects, and other little marine animals.

THE SEA HORSE

Is a small fish of a curious shape. The length seldom reaches twelve inches; the head bears some resemblance to that of a horse, whence originates its name. A long back fin runs from the head to the tail, which is spirally covered. The eggs of this fish are hatched in a pouch, formed by an expansion of the skin, which in some is placed under the belly, and in others at the base of the tail, and which opens to allow the young to get out.

In Europe this fish is often seen in cabinets and museums in a dried state.

CHAPTER IX.

Of Shell Fish in general...The Crustaceous Kind...The Lob-
ster...The Spiny Lobster...The Crab...*The Land Crab...
The Violet Crab...The Soldier Crab...The Shrimp...The
Prawn...The* Tortoise...*The Land Tortoise...The Tur-
tle...Of Testaceous Fishes...Of the* Turbinated, *or Snail
Kind...The Garden Snail...The Fresh-water Snail...The
Sea Snail...The Nautilus...Of* Bivalved *Fish...The Oys-
ter...The Cockle...The Scallop...The Razor Fish..Of Pearls,
and the Fishery...Of* Multivalve *Shell Fish...The Sea
Urchin...The Pholades.*

There are two classes of animals, inhabiting the water,
which commonly receive the name of fishes, entirely different from those we have been describing, and also very
distinct from each other. These are divided by naturalists
into Crustaceous and Testaceous animals: both, totally
unlike fishes in appearance, seem to invert the order of
nature; and as those have their bones on the inside, and
their muscles hung upon them for the purposes of life and
motion, these, on the contrary, have all their bony parts on
the outside, and all their muscles within. Not to talk
mysteriously—all who have seen a lobster or an oyster,
perceive that the shell in these bears a strong analogy to
the bones of other animals; and that, by these shells, the
animal is sustained and defended.

Crustaceous fish, such as the crab and the lobster, have
a shell not quite of a stony hardness, but rather resembling a firm crust, and in some measure capable of yielding.
Testaceous fishes, such as the oyster or cockle, are furnished with a shell of a stony hardness; very brittle, and
incapable of yielding. Of the crustaceous kinds are the

lobster, the crab, and the tortoise : of the testaceous, that numerous tribe of oysters, muscles, cockles, and sea snails, which offer such infinite variety.

THE LOBSTER.

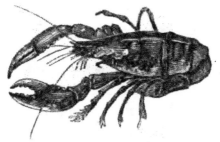

However different in figure the Lobster and the crab may seem, their manners and conformation are nearly the same. With all the voracious appetites of fishes, they are condemned to lead an insect life at the bottom of the water; and though pressed by continual hunger, they are often obliged to wait till accident brings them their prey. Though without any warmth in their bodies, or even red blood circulating through their veins, they are animals wonderfully voracious. Whatever they seize upon that has life, is sure to perish, though ever so well defended : they even devour each other; and, to increase our surprise still more, they may, in some measure, be said to eat themselves; as they change their shell and their stomach every year, and their old stomach is generally the first morsel that serves to glut the new.

The Lobster is an animal of so extraordinary a form, that those who first see it are apt to mistake the head for the tail; but it is soon discovered that the animal moves with its claws foremost; and that the part which plays within itself by joints, like a coat of armour, is the tail. The mouth, like that of insects, opens the long way of the body, not crossways, as with man, and the higher race of ani-

14

mals. It is furnished with two teeth in the mouth, for the comminution of its food; but as these are not sufficient, it has three more in the stomach; one on each side, and the other below. Between the two teeth there is a fleshy substance, in the shape of a tongue. The intestines consist of one long bowel, which reaches from the mouth to the vent; but what this animal differs in from all others is, that the spinal marrow is in the breast bone. It is furnished with two long feelers or horns, that issue on each side of the head, that seem to correct the dimness of the sight, and apprize the animal of its danger, or of its prey. The tail, or that jointed instrument at the other end, is the grand instrument of motion; and with this it can raise itself in the water. Under this we usually see lodged the spawn in great abundance; every pea adhering to the next by a very fine filament, which is scarcely perceivable. Every Lobster is an hermaphrodite, and is supposed to be self impregnated. The ovary, or place where the spawn is first produced, is backwards, towards the tail, where a red substance is always found, and which is nothing but a cluster of peas, that are yet too small for exclusion. From this receptacle there go two canals, that open on each side at the jointures of the shell, at the belly; and through these passages the peas descend to be excluded. and placed under the tail, where the animal preserves them from danger for some time, until they come to maturity. They are then dropped in the sand, where they are soon hatched. Between twelve and thirteen thousand eggs have been counted in one Lobster.

When the young Lobsters are produced, they immediately seek for refuge in the smallest clefts of rocks, and in such like crevices at the bottom of the sea, where the entrance is but small, and the opening can be easily defended. There, without seeming to take any food, they grow

larger in a few weeks time, from the mere accidental substances which the water washes to their retreats. By this
time also they acquire a hard, firm shell, which furnishes
them with both offensive and defensive armour. They
then begin to issue from their fortresses, and boldly creep
along the bottom, in hopes of meeting with diminutive
plunder. The spawn of fish, the smaller animals of their
own kind, but chiefly the worms that keep at the bottom
of the sea, supply them with plenty. They keep in this
manner close among the rocks, busily employed in scratching up the sand with their claws for worms, or surprising
such heedless animals as fall within their grasp: thus they
have little to apprehend, except from each other; for in
them, as among fishes, the large are the most formidable
of all enemies to the small.

But this life of abundance and security is soon to have
a most dangerous interruption ; for the body of the Lobster
still continuing to increase, while its shell remains unalter-
ably the same, the animal becomes too large for its habitation, and imprisoned within the crust that has naturally
gathered round it, there comes on a necessity of getting
free. The young of this kind, therefore, that grow faster,
as we are assured by the fishermen, change their shell oftener than the old, who come to their full growth, and who
remain in the same shell often for two years together. In
general, however, all these animals change their shell once
a year ; and this is not only a most painful operation, but
also subjects them to every danger. Just before casting
its shell, it throws itself upon its back, strikes its claws upon each other, and every limb seems to tremble ; its
feelers are agitated, and the whole body is in violent motion : it then swells itself in an unusual manner, and at
last the shell is seen beginning to divide at its junctures.
It also seems turned inside out; and its stomach comes

15

away with its shell. After this, by the same operation, it disengages itself of the claws, which burst at the joints; the animal, with a tremulous motion casting them off as a man would kick off a boot that was too big for him.

Thus, in a short time, this wonderful creature finds itself at liberty; but in such a weak and enfeebled state, that it continues for several hours motionless. Indeed, so violent and painful is the operation, that many of them die under it; and those which survive are in such a weakly state for some time, that they neither take food nor venture from their retreats. Immediately after this change, they have not only the softness, but the timidity of a worm. Every animal of the deep is then a powerful enemy, which they can neither escape nor oppose; and this, in fact, is the time when the dogfish, the cod, and the ray, devour them by hundreds. But this state of defenceless imbecility continues for a very short time: the animal, in less than two days is seen to have the skin that covered its body grown almost as hard as before; its appetite is seen to increase; and, strange to behold! the first object that tempts its gluttony, is its own stomach, which it so lately was disengaged from. This it devours with great eagerness; and, some time after, eats even its former shell. In about fortyeight hours, in proportion to the animal's health and strength, the new shell is perfectly formed, and as hard as that which was but just thrown aside.

When the Lobster is completely equipped in its new shell, it then appears how much it has grown in the space of a very few days; the dimensions of the old shell being compared with those of the new, it will be found that the creature is increased above a third in its size; and, like a boy that has outgrown his clothes, it seems wonderful how the deserted shell was able to contain so great an animal as entirely fills up the new.

The creature thus furnished, not only with a complete covering, but also a greater share of strength and courage, ventures more boldly among the animals at the bottom; and not a week passes that in its combats it does not suffer some mutilation. A joint, or even a whole claw, is sometimes snapped off in these encounters. At certain seasons of the year, these animals never meet each other without an engagement. In these, to come off with the loss of a leg, or even a claw, is considered as no great calamity; the victor carries off the spoil to feast upon at his leisure, while the other retires from the defeat to wait for a thorough repair. This repair is not long in procuring. From the place where the joint of the claw was cut away, is seen in a most surprising manner to burgeon out the beginning of a new claw. This, if observed, at first, is small and tender, but grows, in the space of three weeks, to be almost as large and as powerful as the old one. We say almost as large, for it never arrives to the full size; and this is the reason we generally find the claws of the Lobsters of unequal magnitude.

Of this extraordinary yet well known animal, there are many varieties, with some differences in the claws, but little in the habits or conformation. It is found above three feet long; and, if we may admit the shrimp and the prawn into the class, though unfurnished with claws, it is seen not above an inch. These all live in the water, and can bear its absence for but a few hours. The shell is black when taken out of the water, but turns red by boiling. The most common way of taking the Lobster is in a basket, or pot, as the fishermen call it, made of wicker work, in which they put the bait, and then throw it to the bottom of the sea, in six or ten fathom water. The Lobsters creep into this for the sake of the bait, but are not able to get out again. The River Crawfish differs little from the

Lobster, but that the one will live only in fresh water, and the other will thrive only in the sea.

The SPINY LOBSTER also differs merely by the offensive armour which it bears upon its back and claws.

THE CRAB.

As the Crab is found upon land as well as in the water, the peculiarity of its situation produces a difference in its habitudes, which it is proper to describe.

The COMMON or BLACK-CLAWED CRAB has three notches on the front; five serrated teeth on each side; the claws elevated; the next joint toothed; the hind feet subulated: the colour a dirty green, but red when boiled. It inhabits almost all shores, and lurks and burrows under the sand.

It changes its shell once a year, generally between Christmas and Easter, and while it is performing this operation it retires among the cavities of rocks, and under large stones. The Crab is an exceedingly quarrelsome animal, and when it has seized on its antagonist it is not easily compelled to forego its hold. In this situation, the captive has no resource but to relinquish the limb, and nature has provided it with the power of accomplishing this in a very curious manner. It stretches the claw out steady, the claw suddenly gives a gentle crack, and the wounded limb drops off, not, as we might be led to infer from reasoning, at the joint, but in the smoothest part.

The LAND CRAB is found in some of the warmer regions of Europe, and in great abundance in all the tropical climates in Africa and America. They are of various kinds, and endued with various properties; some being healthful, delicious, and nourishing food; others poisonous or malignant to the last degree; some are not above half an inch broad, others are found a foot over; some are of a dirty brown, and others beautifully mottled. The animal called the VIOLET CRAB of the Caribee Islands is the most noted, both for its shape, the delicacy of its flesh, and the singularity of its manners.

The Violet Crab somewhat resembles two hands cut through the middle and joined together; for each side looks like four fingers, and the two nippers or claws resemble the thumbs. All the rest of the body is covered with a shell as large as a man's hand, and bunched in the middle, on the fore part of which there are two long eyes of the size of a grain of barley, as transparent as crystal and as hard as horn. A little below these is the mouth, covered with a sort of barbs, under which there are two broad, sharp teeth, as white as snow. They are not placed, as in other animals, crossways, but in the opposite direction, not much unlike the blade of a pair of scissors. With these teeth they can easily cut leaves, fruits, and rotten wood, which is their usual food. But their principal instrument for cutting and seizing their food is their nippers, which catch such a hold, that the animal 'loses the limb sooner than its grasp, and is often seen scampering off, having left its claw still holding fast upon the enemy. The faithful claw seems to perform its duty, and keeps for above a minute fastened upon the finger while the Crab is making off.* In fact, it loses no great matter by leaving a leg or an

* Brown's Jamaica, p. 428.

arm, for they soon grow again, and the animal is found as
perfect as before.

This, however, is the least surprising part of this creature's history: and what we are going to relate, were it
not as well known and as confidently confirmed as any
other circumstance in natural history, might well stagger
our belief. These animals live not only in a kind of society in their retreats in the mountains, but regularly once a
year march down to the sea side in a body of some millions
at a time. As they multiply in great numbers, they choose
the months of April or May to begin their expedition; and
then sally out by thousands from the stumps of hollow
trees, from the clifts of rocks, and from the holes which
they dig for themselves under the surface of the earth.
At that time the whole ground is covered with this band
of adventurers; there is no setting down one's foot without treading upon them. The sea is the place of destina-
tion, and to that they direct their march with right lined
precision. No geometrician could send them to their des-
tination by a shorter course; they neither turn to the right
nor left, whatever obstacles intervene; and even if they
meet with a house, they will attempt to scale the walls to
keep the unbroken tenor of their way. But though
this be the general order of their route, they upon other
occasions are compelled to conform to the face of the
country; and if it be intersected by rivers, they are then
seen to wind along the course of the stream. The proces-
sion sets forward from the mountains with the regularity
of an army under the guidance of an experienced com-
mander. They are commonly divided into three battalions:
of which the first consists of the strongest and boldest
males, that, like pioneers, march forward to clear the route
and face the greatest dangers. These are often obliged
to halt for want of rain, and go into the most convenient

encampment till the weather changes. The main body of
the army is composed of females, which never leave the
mountains till the rain is set in for some time, and then
descend in regular battalia, being formed into columns of
fifty paces broad, and three miles deep, and so close that
they almost cover the ground. Three or four days
after this the rear-guard follows; a straggling, undisci-
plined tribe, consisting of males and females, but neither so robust nor so numerous as the former. The night
is their chief time of proceeding; but if it rains by
day, they do not fail to profit by the occasion; and they
continue to move forward in their slow uniform manner. When the sun shines, and is hot upon the surface
of the ground, they then make a universal halt, and wait
till the cool of the evening. When they are terrified,
they march back in a confused disorderly manner, holding
up their nippers, with which they sometimes tear off a
piece of the skin, and then leave the weapon where they
inflicted the wound. They even try to intimidate their
enemies; for they often clatter their nippers together, as if
it were to threaten those that come to disturb them. But
though they thus strive to be formidable to man, they are
much more so to each other; for they are possessed of one
most unsocial property, which is, that if any of them by
accident is maimed in such a manner as to be incapable of
proceeding, the rest fall upon and devour it on the spot, and
then pursue their journey.

When, after a fatiguing march, and escaping a thousand dangers, for they are sometimes three months in getting to the shore, they have arrived at their destined port,
they prepare to cast their spawn. The peas are yet within their bodies, and not excluded, as is usual in animals of
this kind, under the tail; for the creature waits for the
benefit of the sea water to help the delivery. For this

purpose, the Crab has no sooner reached the shore than it eagerly goes to the edge of the water, and lets the waves wash over its body two or three times. This seems only a preparation for bringing their spawn to maturity; for without farther delay they withdraw to seek a lodging upon land: in the mean time, the spawn grows larger, is excluded out of the body, and sticks to the barbs under the flab, or more properly the tail. This bunch is seen as big as a hen's egg, and exactly resembling the roes of herrings. In this state of pregnancy, they once more seek the shore for the last time, and shaking off their spawn into the water, leave accident to bring it to maturity. At this time whole shoals of hungry fish are at the shore, and about two thirds of the Crabs' eggs are immediately devoured by these rapacious invaders. The eggs that escape are hatched under the sand; and soon after millions at a time of these little Crabs are seen quitting the shore, and slowly travelling up to the mountains.

The old ones, however, are not so active to return; they have become so feeble and lean, that they can hardly creep along, and the flesh at that time changes its colour. Most of them, therefore, are obliged to continue in the flat parts of the country till they recover, making holes in the earth, which they cover at the mouth with leaves and dirt, so that no air may enter. There they throw off their old shells, which they leave as it were quite whole, the place where they are opened on the belly being unseen. At that time they are quite naked, and almost without motion for six days together, when they become so fat as to be delicious food. They have then under their stomachs four large white stones, which gradually decrease in proportion as the shell hardens, and when they come to perfection are not to be found. It is at that time that the animal is seen slowly making its way back; and all this is most commonly performed in the space of six weeks.

The descent of these creatures for such important purposes deserves our admiration; but there is an animal of the lobster kind that annually descends from its mountains in like manner, and for purposes still more important and various. Its descent is not only to produce an offspring, but to provide itself a covering; not only to secure a family, but to furnish a house. The animal in question is the SOLDIER CRAB, or HERMIT CRAB, which has some simili-

tude to the lobster, if divested of its shell. It is usually about four inches long, has no shell behind, but is covered down to the tail with a rough skin, terminating in a point. It is, however, armed with strong hard nippers before, like the lobster; and one of them is as thick as a man's thumb, and pinches most powerfully. It is, as was said, without a shell to any part except its nippers; but what Nature has denied this animal it takes care to supply by art; and taking possession of the deserted shell of some other animal, it resides in it, till, by growing too large for its habitation, it is under a necessity of change. It is a native of the West India Islands; and like the former, it is seen every year descending from the mountains to the seashore, to deposit its spawn, and to provide itself with a new shell. This is a most bustling time with it, having so many things to do; and, in fact, very busy it appears. It is very probable that its first care is to provide for its offspring before it attends to its own wants; and it is thought, from the number of little shells which it is seen examining, that it deposits its spawn in them, which thus is placed in perfect security till the time of exclusion.

It is very diverting to observe these animals when changing the shell. The little Soldier is seen busily parading the shore along that line of pebbles and shells which is formed by the extremest wave, still, however, dragging its old incommodious habitation at its tail, unwilling to part with one shell, even though a troublesome appendage, till it can find one more convenient. It is seen stopping at one shell, turning it, and passing it by; going on to another, contemplating that for a while, and then slipping its tail from its old habitation to try on the new; this also is found to be inconvenient, and it quickly returns to its old shell again. In this manner it frequently changes, till at last it finds one light, roomy, and commodious; to this it adheres, though the shell be sometimes so large as to hide the body of the animal, claws and all.

Yet it is not only till after many trials, but many combats also, that the Soldier is completely equipped; for there is often a contest between two of them for some well-looking favourite shell for which they are rivals. They both endeavour to take possession: they strike with their claws; they bite each other, till the weakest is obliged to yield, by giving up the object of dispute. It is then that the victor immediately takes possession, and parades in his new conquest three or four times backward and forward upon the strand before his envious antagonist.

When this animal is taken, it sends forth a feeble cry, endeavouring to seize the enemy with its nippers; which if it fastens upon, it will sooner die than quit the grasp. The wound is very painful, and not easily cured. For this reason, and as it is not much esteemed for its flesh, it is generally permitted to return to its old retreat to the mountains in safety. There it continues till the necessity of changing once more, and the desire of producing an offspring, expose it to fresh dangers the year ensuing.

There are many other species of this animal, such as the LOBSTER CRAB, the RIVER CRAB, the MINUTE CRAB, which is found in the inside of muscles, &c.

THE SHRIMP AND PRAWN.

SHRIMPS possess long slender feelers, and between them two thin projecting laminæ; the claws have a single hooked moveable fang; they have three pair of legs; seven joints in the tail; the middle caudal fin subulated, the four others round and fringed; a spine on the exterior side of

each of the outmost. These animals, which are of a delicate flavour, inhabit the shores of Britain in vast quantities, ascend the rivers, and even find their way into the ditches and ponds of salt marshes. Those caught in the sea are, however, much better than those of the rivers. They are also found in the United States.

The Prawn is not unlike the Shrimp, but exceeds it in size, being at least three times as big; and is more pleasing in color, it having, when boiled, the most beautiful pink tint all over its body. The flesh is better tasted than that of the Shrimp. It has a long horn in front of its head, compressed vertically, which bends somewhat upward, and is serrated both above and below. Seaweeds, and the vicinity of rocks near the shore, are the haunts of the Prawn, which, unlike the Shrimp, seldom enters the mouths of rivers. It usually swims on its back, but when in danger it throws itself on one side, and springs backward to a considerable distance.

THE TORTOISE.

Tortoises are usually divided into those that live upon land, and those that subsist in the water; and use has made a distinction even in the name, the one being called Tortoises, the other Turtles. However, Seba has proved that all Tortoises are amphibious; that the land Tortoise will live in the water; and that the sea Turtle can be fed upon land. A land Tortoise was brought to him, that was caught in one of the canals of Amsterdam, which he kept for half a year in his house, where it lived very well contented in both elements. When in the water, it remained with its head above the surface; when placed in the sun, it seemed delighted with its beams, and continued immoveable while it felt their warmth. The difference, therefore, in these animals arises rather from their habits than their conformation; and, upon examination, there will be less variety found between them than between birds that live upon land and those that swim upon the water.

All Tortoises, in their external form, much resemble each other; their outward covering being composed of two great shells, the one laid upon the other, and only touching at the edges; however, when we come to look closer, we shall find that the upper shell is composed of no less than thirteen pieces. There are two holes at either edge of this vaulted body; one for a very small head, shoulders, and arms, to peep through; the other at the opposite edge, for the feet and the tail. These shells the animal is never disengaged from; and they serve for its defence against every creature but man.

THE LAND TORTOISE

Is generally found from one to five feet long, from the end of the snout to the end of the tail; and from five inches to a foot and a half across the back. It has a small head, somewhat resembling that of a serpent; an eye without the upper lid; the under eyelid serving to cover and keep that organ in safety. It has a strong, scaly tail, like the lizard. Its head the animal can put out and hide, at pleasure, under the great penthouse of its shell; there it can remain secure from all attacks. As the tortoise lives wholly upon vegetable food, it never seeks the encounter; yet if any of the smaller animals attempt to invade its repose, they are sure to suffer. The Tortoise, impregnably defended, is furnished with such a strength of jaw, that, though armed only with bony plates instead of teeth, wherever it fastens, it infallibly keeps its hold until it has taken out the piece.

Though peaceable in itself, it is formed for war in another respect, for it seems almost endued with immortality. Nothing can kill it; the depriving it of one of its members is but a slight injury; it will live, though deprived of the brain; it will live, though deprived of its head. Tortoises are commonly known to exceed eighty years; and there was one kept in the Archbishop of Canterbury's garden at Lambeth, London, that was remembered above a hundred and twenty. It was at last killed by the severity of the frost, from which it had not sufficiently defended itself in its winter retreat, which was a heap of sand at the bottom of the garden.

Though there is a circulation of blood in the Tortoise, yet, as the lungs are left out of the circulation, the animal is capable of continuing to live without continuing to breathe. In this it resembles the bat, the serpent, the mole, and the lizard; like them it takes up its dark residence for the winter, and, at that time, when its food is no longer in plenty, it happily becomes insensible to the want. But it must not be supposed that, while it is thus at rest, it totally discontinues to breathe; on the contrary, an animal of this kind, if put into a close vessel, without air, will soon be stifled; though not so readily as in a state of vigour and activity.

The eggs of all the Tortoise kind, like those of birds, are furnished with a yolk and a white; but the shell is different, being somewhat like those soft eggs that hens exclude before their time: however, this shell is much thicker and stronger, and is a longer time in coming to maturity in the womb. The Land Tortoise lays but a few in number, if compared to the sea Turtle, who deposits from a hundred and fifty to two hundred in a season.

The amount of the Land Tortoise's eggs we have not been able to learn; but, from the scarceness of the animal, we are apt to think they cannot be very numerous. When it prepares to lay, the female scratches a slight depression in the earth, generally in a warm situation, where the beams of the sun have their full effect. There depositing her eggs, and covering them with grass and leaves, she forsakes them, to be hatched by the heat of the season. The young Tortoises are generally excluded in about twenty-six days; but, as the heat of the weather assists, or its coldness retards, incubation, sometimes it happens that there is a difference of two or three days. The little animals no sooner leave the egg, than they seek for their provision, entirely self-taught; and their shell, with which they are covered from the beginning, expands and grows larger with age. As it is composed of a variety of pieces, they are all capable of extension at their sutures; and the shell admits of increase in every direction.

It is common enough to take these animals into gardens, as they are thought to destroy insects and snails in great abundance. We are even told that, in hot countries, they are admitted into a domestic state, as they are great destroyers of bugs.

THE SEA TORTOISE, OR TURTLE,

As it is now called, is generally found larger than the former.

The *Great Mediterranean Turtle* is the largest of the Turtle kind with which we are acquainted. It is found from five to eight feet long, and from six to nine hundred pounds weight; but, unluckily, its utility bears no proportion to its size, as it is unfit for food, and sometimes poisons those who eat it. The shell also, which is a tough, strong integument, resembling a hide, is unfit for all serviceable purposes. One of these animals was taken in the year 1729, at the mouth of the Loire, in France, in nets that were not designed for so large a capture. This Turtle, which was of enormous strength, by its own struggles involved itself in the nets in such a manner as to be incapable of doing mischief: yet, even thus shackled, it appeared terrible to the fishermen, who were at first for flying; but, finding it impotent, they gathered courage to drag it on shore, where it made a most horrible bellowing; and when they began to knock it on the head with their gaffs, it was to be heard at half a mile's distance. They were still farther intimidated by its nauseous and pestilential breath, which so powerfully affected them that they were near fainting. This animal wanted but four inches of being eight feet long, and was about two feet over: its shell more resembled leather than the shell of a Tortoise; and unlike all other animals of this kind, it was furnished with teeth in each jaw, one rank behind another, like those of a shark: its feet also, different from the rest of this kind, wanted claws; and the tail was quite disengaged from the shell, and fifteen inches long, more resembling that of a quadruped than a Tortoise.

These are a formidable and useless kind, if compared to the Turtle caught in the South Seas and the Indian Ocean.

These are of different kinds; not only unlike each other in form, but furnishing man with very different advantages. They are usually distinguished by sailors into four kinds; the *Trunk Turtle*, the *Loggerhead*, the *Hawksbill*, and the *Green Turtle*.

The *Hawksbill* or *Imbricated Turtle* is the least of the four, and has a long and small mouth, somewhat resembling the bill of a hawk. The flesh of this also is very indifferent eating; but the shell serves for the most valuable purposes. This is the animal that supplies the tortoise-shell, of which such a variety of beautiful trinkets are made.

But of all animals of the Tortoise kind, the green Turtle

is the most noted and the most valuable, from the delicacy of its flesh, and its nutritive qualities, together with the property of being easily digested. It is generally found about two hundred weight; though some are five hundred, and others not above fifty. Dampier mentions one so large that a boy of ten years of age, the son of Captain Rock, went from the shore in the shell of it, as a boat, to his father's ship.

This animal seldom comes from the sea but to deposit its eggs. Its chief food consists of the mangrove, the blackwood tree, and other marine plants. When the weather is fair, the Turtles are sometimes seen feeding in great numbers, like flocks of sheep, several fathoms deep upon the verdant carpet below. They frequent the creeks and shallows, where they are usually taken, but they are extremely shy of boats and men, and swim remarkably

fast. They are sometimes found on the shore at night; and numbers of them are thus captured by the sailors.

When the time for laying approaches, the female is seen, towards the setting of the sun, drawing near the shore, and looking earnestly about her, as if afraid of being discovered. When she perceives any person on shore, she seeks for another place; but if otherwise, she lands when it is dark, and goes to take a survey of the sand where she designs to lay. Having marked the spot, she goes back without laying, for that night, to the ocean again; but the next night returns to deposit a part of her burthen. She begins by working and digging in the sand with her fore feet, till she has made a round hole, a foot broad and a foot and a half deep, just at the place a little above where the water reaches highest. This done, she lays eighty or ninety eggs at a time, each as big as a pigeon's egg. The eggs are covered with a tough white skin, like wetted parchment. When she has done laying, she covers the hole so dexterously, that it is no easy matter to find the place. When the Turtle has done laying, she returns to the sea, and leaves her eggs to be hatched by the heat of the sun. At the end of fifteen days, she lays about the same number of eggs again; and at the end of another fifteen days, she repeats the same; three times in all, using the same precautions every time for their safety.

In about twenty-four or twenty-five days after laying, the eggs are hatched by the heat of the sun; and the young Turtles, being about as big as quails, are seen bursting from the sand, as if earth-born, and running directly to the sea, with instinct only for their guide; but, to their great misfortune, it often happens that their strength being small, the surges of the sea, for some few days, beat them back upon the shore. Thus exposed, they remain a prey to thousands of birds that haunt the coast; and these, swooping down upon them, carry off the greatest part, and sometimes the whole brood, before they have strength sufficient to withstand the waves or dive to the bottom. Tur-

tles 'are generally caught in two ways, by nets, and
by what is called pegging. The peg is of iron, and something larger than a ten-penny nail, and without a barb; to
this is affixed an iron socket, in which is inserted a long pole,
and the peg is held by a tolerably strong line. When the
Turtle is struck, the hunter disengages the pole, and draws
the Turtle to the boat by the line. They are also caught
by turning them on their back while they are asleep.

There is yet another way which, though seemingly awkward, is said to be attended with very great success. A
good diver places himself at the head of the boat; and
when the Turtles are observed, which they sometimes are
in great numbers, asleep on the surface, he immediately
quits the vessel at about fifty yards distance, and, keeping
still under water, directs his passage to where the Turtle
was seen, and, coming up beneath, seizes it by the hind

fin; the animal, awaking, struggles to get free; and by
this both are kept at the surface until the boat arrives to

take them in. The natives of the islands in the Pacific
Ocean often catch these animals by swimming into the
water, and turning them on their backs; they then tow
them ashore,

The shell of TESTACEOUS FISHES may be considered as
a habitation supplied by nature. It is a hard stony substance, made by some in the manner of a wall. Part of
the stony substance the animal derives from outward objects, and the fluids of the animal itself furnish the cement.
These united make that firm covering which shell fish
generally reside in till they die.

But, in order to give a more exact idea of the manner
in which sea-shells are formed, we must have recourse to
an animal that lives upon land, with the formation of whose
shell we are best acquainted. This is the garden snail,
that carries its box upon its back.

To begin with the animal in its earliest state, and trace
the progress of its shell from the time it first appears—
The instant the young snail leaves the egg, it carries its
shell or its box on its back. It does not leave the egg till
it is arrived at a certain growth, when its little habitation
is sufficiently hardened. This beginning of the shell is not
much bigger than a pin's head, but grows in a very rapid
manner, having at first but two circumvolutions, for the
rest are added as the snail grows larger. In proportion as
the animal increases in size, the circumvolutions of the
shell increase also, until the number of those volutes come
to be five, which is never exceeded.

The part where the animal enlarges its shell is at the
mouth, to which it adds in proportion as it finds itself stint-
ed in its habitation below. Being about to enlarge its
shell, it is seen with its little teeth biting and clearing
away the scaly skin that grows at the edges. It is sometimes seen to eat those bits it thus takes off; and at other

times it only cleans away the margin when covered with films, and then adds another rim to its shell.

For the purposes of making the shell, which is natural to the animal, and without which it would not live three days, its whole body is furnished with glands, from the orifices of which flows out a kind of slimy fluid, like small spiders' threads, which join together in one common crust or surface, and in time condense and acquire a stony hardness. It is this slimy humor that grows into a membrane, and afterwards a stony skin, nor can it have escaped any who has observed the track of a snail; that glistening substance which it leaves on the floor or the wall is no other than the materials with which the animal adds to its shell, or repairs it when broken.

With respect to the figure of shells, Aristotle has divided them into three kinds; and his method is, above all others, the most conformable to nature. These are, first, the *Univalve* or *Turbinated*, which consist of one piece, like the box of a snail; secondly, the *Bivalve*, consisting of two pieces united by a hinge, like an óyster; and, thirdly, the *Multivalve*, consisting of more than two pieces, as the acorn-shell, which has not less than twelve pieces that go to its composition. All these kinds are found in the sea at different depths, and are valuable in proportion to their scarceness or beauty. All shells are formed of an animal or calcareous earth, that ferments with vinegar and other acids, and that burns into lime, and will not easily melt into glass.

Every shell, wherever it is found, is the spoil of some animal that once found shelter therein. It matters not by what unaccountable means they may have wandered from the sea; but they exhibit all and the most certain marks of their origin. From their numbers and situation, we are led to conjecture, that the sea reached the places where they are found; and from their varieties we learn how little we know of all the sea contains at present; as the earth furnishes many kinds which our most exact and industrious shell collectors have not been able to fish up from the deep.

UNIVALVE OR TURBINATED SHELL FISH.

To conceive the manner in which these animals subsist that are hid from us at the bottom of the deep, we must again have recourse to one of a similar nature and formation, that we know, viz. the GARDEN SNAIL. It is furnished

with the organs of life in a manner almost as complete as the largest animal; with a tongue, brain, salival ducts, glands, nerves, stomach and intestines, liver, heart, and blood-vessels: besides these, it has a purple bag that furnishes a red matter to different parts of the body, together with strong muscles that hold it to the shell, and which are hardened, like tendons, at their insertion.

But these it possesses in common with other animals. We must now see what it has peculiar to itself. The first striking peculiarity is, that the animal has got its eyes on the points of its largest horns. When the snail is in motion, four horns are distinctly seen; but the two uppermost and longest deserve peculiar consideration, both on account of the various motions with which they are endued, as well as their having their eyes fixed at the extreme ends of them. The eyes the animal can direct to different objects at pleasure, by a regular motion out of the body; and sometimes it hides them, by a very swift contraction into the belly. Under the small horns is the animal's mouth; and though it may appear too soft a substance to be furnished with teeth, yet it has not less than eight of them, with which it devours leaves, and other substances, seemingly harder than itself; and with which it sometimes bites off pieces of its own shell.

At the expiration of eighteen days after coupling, the snails produce their eggs, and hide them in the earth with the greatest solicitude and industry. These eggs are in great numbers, round, white, and covered with a soft shell: they are also stuck to each other by an imperceptible slime; like a bunch of grapes, of about the size of a small pea.

The snail is possessed not only of a power of retreating into its shell, but of mending it when broken. Sometimes these animals are crushed seemingly to pieces; and to all appearance utterly destroyed; yet still they set themselves to work, and, in a few days, mend all their numerous breaches. The same substance by which the shell is originally made, goes to the re-establishment of the ruined habitation.

As the Snail is furnished with all the organs of life and sensation, it is not wonderful to see it very voracious. It chiefly subsists upon the leaves of plants and trees; but is very delicate in its choice. At the approach of winter, it buries itself in the earth, or retires to some hole to continue in a torpid state, during the severity of the season. It is sometimes seen alone; but more frequently in company in its retreat; several being usually found together apparently deprived of life and sensation. For the purpose of continuing in greater warmth and security, the snail forms a cover or lid to the mouth of its shell with its slime, which stops it up entirely, and thus protects it from

every external danger. When the cover is formed too
thick, the snail then breaks a little hole in it, to correct
the effect of that closeness, which proceeded from too
much caution. In this manner, sheltered in its hole from
the weather, defended in its shell by a cover, it sleeps during the winter; and for six or seven months, continues
without food or motion, until the genial call of spring
breaks its slumber, and excites its activity.

The Snail, having slept for so long a season, awakes in
one of the first fine days of April, breaks open its cell,
and sallies forth to seek for nourishment. At first, it is
not very difficult in the choice of its food; almost any vegetable that is green seems welcome; but the succulent
plants of the garden are chiefly grateful; and the various
kinds of pulse are, at some seasons, almost wholly destroyed by their numbers. A wet season is generally favourable to their production; for this animal cannot bear very
dry seasons, or dry places, as they cause too great a consumption of its slime, without plenty of which it cannot
subsist in health and vigour.

Such are the most striking particulars in the history of
this animal; and this may serve as a general picture, to
which the manners and habitudes of the other tribes of
this class may be compared and referred. These are the
Sea Snail, of which naturalists have, from the apparent difference of their shells, mentioned fifteen kinds;* the Fresh
Water Snail, of which there are eight kinds; and the Land
Snail, of which there are five; and these all bear a strong
resemblance to the Garden Snail. All *Snails that live in
water*, are peculiarly furnished with a contrivance by nature, for rising to the surface, or sinking to the bottom.
The manner in which this is performed is by opening and

* D'Argenville's Conchyliologie.

shutting an orifice on the right side of the neck, which is furnished with muscles for that purpose. The Snail sometimes gathers this aperture into an oblong tube, and stretches or protends it above the surface of the water, in order to draw in or expel the air, as it finds occasion. This may not only be seen, but heard also by the noise which the Snail makes in moving the water. By dilating this, it rises; by compressing it, the animal sinks to the bottom.

But what renders these animals far more worthy of notice is, that they are viviparous, and bring forth their young not only alive, but with their shells upon their backs. This seems surprising; yet it is incontestably true: the young come to some degree of perfection in the womb of the parent; there they receive their stony coat; and thence are excluded, with a complete apparatus for subsistence.

This striking difference between the Fresh-Water and the Garden Snail obtains also in some of the SEA KIND; among which there are some that are found viviparous, while others lay eggs in the usual manner. But this is not the only difference between Land and Sea Snails. Many of the latter entirely want horns; and none of them have above two. Indeed, if the horns of Snails be furnished with eyes, and if, as some are willing to think, the length of the horn, like the tube of a telescope, assists vision, these animals, that chiefly reside in the gloomy bottom of the deep, can have no great occasion for them. Eyes would be unnecessary to creatures whose food is usually concealed in the darkest places; and who, possessed of very little motion, are obliged to grope for what they subsist on. To such, eyes would rather be an obstruction than an advantage; and perhaps even those that live upon land are without them!

There is a difference also in the position of the mouth, in the Garden and the Water Snail. In the former, the

x5

mouth is placed crosswise, as in quadrupeds; furnished with jawbones, lips, and teeth. In most of the Sea Snails, the mouth is placed longitudinally in the head; and, in some, obliquely, or on one side. Others, of the trochus kind, have no mouth whatsoever; but are furnished with a trunk, very long in some kinds, and shorter in others.

Of all Sea Snails, that which is most frequently seen swimming upon the surface, and whose shell is the thinnest and most easily pierced, is the NAUTILUS. Whether, upon these occasions, it is employed in escaping its numerous enemies at the bottom, or seeking for food at the surface, we will not venture to decide. It seems most probable, that the former is the cause of its frequently appearing; for, upon opening the stomach, it is found to contain chiefly that food which it finds at the bottom.

Although there are several species of the Nautilus, yet they all may be divided into two: the one with a white shell, as thin as paper, which it often is seen to quit, and again to resume; the other with a thicker shell, sometimes of a beautiful mother-of-pearl colour, and that quits its shell but rarely. This shell outwardly resembles that of a large snail, but is generally six or eight inches across: within, it is divided into forty partitions, that communicate with each other by doors, if we may so call them, through which one could not thrust a goose quill: almost the whole internal part of the shell is filled by the animal, the body of which, like its habitation, is divided into as many parts as there are chambers in its shell: all the parts of its body communicate with each other, through the doors or openings, by a long blood-vessel, which runs from the head to the tail: thus the body of the animal, if taken out of the shell, may be likened to a number of soft bits of flesh, of which there are forty threaded upon a string. From this extraordinary conformation, one would not be apt to suppose that the Nautilus sometimes quitted its shell, and returned to it again; yet nothing, though seemingly impossible, is more certain. The manner by which it contrives to disengage every part of its body from so intricate a habitation—by which it makes a substance, to appearance as thick as one's wrist, pass through forty doors, each of which would scarcely admit a goose quill—is not yet discovered: but the fact is certain; for the animal is often found without its shell; and the shell more frequently destitute of the animal. It is most probable, that it has a power of making the substance of one section of its body remove up into that which is next; and thus, by multiplied removals, it gets free.

But this, though very strange, is not the peculiarity for which the Nautilus has been the most distinguished. Its spreading the thin oar, and catching the flying gale, to use the poet's description of it, has chiefly excited human curiosity. These animals, particularly those of the white, light kind, are chiefly found in the Mediterranean; and scarce any who have sailed on that sea, but must have often seen them. When the sea is calm, they are observed floating on the surface; some spreading their little sail; some rowing with their feet, as if for life and death; and others still, floating upon their mouths, like a ship with the keel upward. If taken while thus employed, and examined, the extraordinary mechanism of their limbs for sailing will appear more manifest. The Nautilus is furnished with eight feet, which issue near the mouth, and may as properly be called barbs: these are connected to each other by a thin skin, like that between the toes of a duck, but much thinner, and more transparent. Of these eight feet thus connected, six are short, and these are held up as sails to catch the wind in sailing: the two others are longer, and are kept in the water; serving, like paddles,

to steer their course by.   When the weather is quite calm,
and the animal is pursued from below, it is then seen ex-
panding only a part of its sail, and rowing with the rest:
whenever it is interrupted, or fears danger from above, it
instantly furls the sail, catches in all its oars, turns its
shell mouth downward, and instantly sinks to the bottom.
Sometimes also it is seen pumping the water from its
leaking hulk; and, when unfit for sailing, deserts its shell
entirely. The forsaken hulk is seen floating along, till it
dashes, by a kind of shipwreck, upon the rocks or the
shore.

BIVALVE SHELL FISH.

It may seem whimsical to make a distinction between the
animal perfection of turbinated and *Bivalved Shell Fish*, or
to grant a degree of superiority to the snail above the oyster.   Yet this distinction strongly and apparently obtains
in nature; and we shall find .the bivalved tribe of animals
in every respect inferior to those we have been describing.

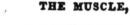

THE MUSCLE,

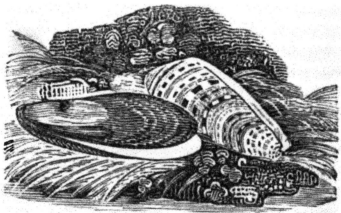

As is well known, whether belonging to fresh or salt water, consists of two equal shells, joined at the back by a

strong muscular ligament that answers all the purposes of
a hinge. By the elastic contraction of this, the animal
can open its shells at pleasure, about a quarter of an inch
from each other. The fish is fixed to either shell by four
tendons, by means of which it shuts them close, and keeps
its body firm from being crushed by any shock against the
walls of its own habitation. It is furnished, like all other
animals of this kind, with vital organs, though these are
situated in a very extraordinary manner. It has a mouth
furnished with two fleshy lips ; its intestines begin at the
bottom of the mouth, pass through the brain, and make a
number of circumvolutions through the liver; on leaving
this organ, they go on straight into the heart, which they
penetrate, and end in the anus; near which the lungs are
placed, and through which it breathes, like those of the
snail kind; and in this manner its languid circulation is
carried on.

The multitude of these animals in some places is very
great; but from their defenceless state, the number of
their destroyers are in equal proportion.

But notwithstanding the number of this creature's animated enemies, it seems still more fearful of the agitations
of the element in which it resides; for if dashed against
rocks, or thrown far on the beach, it is destroyed without
a power of redress. In order to guard against these,
which are to this animal the commonest and the most fatal
accidents, although it has a power of slow motion, which
we shall presently describe, yet it endeavours to become
stationary, and to attach itself to any fixed object it happens to be near. For this purpose it is furnished with a
very singular capacity of binding itself by a number of
threads to whatever object it approaches; and these Reaumur supposed it to spin artificially, as spiders their webs,
which they fasten against a wall. Of this, however, later

philosophers have found very great reason to doubt. It is therefore supposed that these threads, which are usually called the beard of the Muscle, are the natural growth of the animal's body, and by no means produced at pleasure.

Its instrument of motion, by which it contrives to reach the object it wants to bind itself to, is that muscular substance resembling a tongue, which is found long in proportion to the size of the Muscle. In some it is two inches long, in others not a third part of these dimensions. This the animal has a power of thrusting out of its shell; and with this it is capable of making a slight furrow in the sand at the bottom. By means of this furrow it can erect itself upon the edge of its shell; and thus continuing to make the furrow in proportion as it goes forward, it reaches out its tongue, that answers the purpose of an arm, and thus carries its shell edgeways, as in a groove, until it reach the point intended. There where it determines to take up its residence it fixes the ends of its beard, which are glutinous, to the rock or the object, whatever it be; and thus, like a ship at anchor, braves all the agitations of the water. The beards have been seen a foot and a half long; and of this substance the natives of Palermo sometimes make gloves and stockings.

These shell fish are found in lakes, rivers, and in the sea. Those of the lake often grow to a very large size; but they seem a solitary animal, and are found generally separate from each other. Those of rivers are not so large, but in yet greater abundance; but the Sea Muscle is in most plenty. These are often bred artificially in salt water marshes that are overflowed by the tide; the fishermen throwing them in at the proper seasons; and there being undisturbed by the agitations of the sea, and not preyed upon by their powerful enemies at the bottom, they cast their eggs, which soon become perfect animals, and these are generally found in clusters of several dozen together.

It requires a year for the peopling a Muscle-bed; so that, if the number consists of forty thousand, a tenth part may annually be left for the peopling the bed anew. Muscles are taken from their beds from the month of July to October; and they are sold at a very moderate price.

From this animal the OYSTER differs very little, except in the thickness of its shell, and its greater imbecility.

The Oyster, like the muscle, is formed with organs of life and respiration, with intestines which are very voluminous, a liver, lungs, and heart. Like the muscle, it is self-impregnated; and the shell, which the animal soon acquires, serves it for its future habitation. Like the muscle, it opens its shell to receive the influx of water, and like that animal, is strongly attached to its shells both above and below. The Oyster respires by means of gills. The water is drawn in at the mouth, which is a small opening in the upper part of the body, and proceeds thence down a long canal, constituting the base of the gills, and so out again, the animal retaining such a portion of air as is necessary for the functions of the body.

The Oyster differs from the muscle in being utterly unable to change its situation. It is entirely without that tongue which we saw answering the purposes of an arm in the other animal; but nevertheless is often attached very firmly to any object it happens to approach. Nothing is so common in the rivers of the tropical climates as to see Oysters growing even amidst the branches of the forest. Many trees, which grow along the banks of the stream, often bend their branches into the water, and particularly the mangrove, which chiefly delights in a moist situation. To these the Oysters hang in clusters, like apples upon the most fertile tree; and in proportion as the weight of the fish sinks the plant into the water, where it still continues growing, the. number of Oysters increase, and hang upon the branches. This is effected by means of a glue proper to themselves, which, when it cements, the joining is as hard as the shell, and is as difficultly broken.

Oysters usually cast their spawn in May, which at first appears like drops of candle-grease, and sticks to any hard substance it falls upon. These are covered with a shell in two or three days; and in three years the animal is large enough to be brought to market. As they invariably remain in the places where they are laid, and as they grow without any other seeming food than the afflux of sea water, it is the cust..m at Colchester, and other parts of England, where the tide settles in marshes on land, to pick up great quantities of small Oysters along the shore, which, when first gathered, seldom exceed the size of a sixpence. These are deposited in beds where the tide comes in, and in two or three years grow to a tolerable size. They are said to be better tasted for being thus sheltered from the agitation of the deep; and a mixture of fresh water entering into these repositories, is said to improve their flavour, and to increase their growth and fatness. Most of the Oysters sold in Boston are taken in some part of Long Island Sound, and kept a year at Cape Cod, where they grow much larger, and are better than when first taken.

The Oysters, however, which are prepared in this manner, are by no means so large as those found sticking to
rocks at the bottom of the sea, usually called Rock Oysters.
These are sometimes found as broad as a plate, and are
admired by some as excellent food.   But what is the size
of these compared to the Oysters of the East Indies, some
of whose shells we have seen two feet over?   The Oysters
found along the coast of Coromandel are capable of
furnishing a plentiful meal to eight or ten men; but it
seems universally agreed that they are no way comparable
to ours for delicacy or flavour.   The Oysters taken on the
coast of England have a strong taste of copper, which they
derive from the copper banks.   They are at first very disgusting to an American palate.

Thus the muscle and the oyster appear to have but few
distinctions, except in their shape, and the power of motion
in the former.   Other bivalved shell fish, such as the Cockle, the Scallop, and the Razor Shell, have differences
equally minute.   The power of changing place, which
some of them effect in a manner quite peculiar to themselves, makes their greatest difference.

Of the Cardium, or Cockle tribe, there are more than
fifty species; some or other of which are to be procured
on the sandy shores of all the known seas.   They are
mostly found immersed a few inches deep in the sand.   In
size the different species vary considerably, some being
five or six inches in diameter, and others not more than
half an inch.   The Cockle has a tolerable degree of loco-
motive power, in consequence of its triangular yellow foot,
which is conspicuous on the shell being opened.   With
this foot it can also draw into threads its glutinous matter,
and thus in a manner anchor itself on the spot that it has
chosen for its residence.   The opening of the shell is pro-

tected by a soft membrane, which wholly closes up the front, except in two places, at each of which there is a small, yellow, fringed tube. Through these tubes the animal receives and ejects the water which conveys nutriment to its body.

The CARDIUM EDULE, or COMMON COCKLE, which is the species most common in England, has a grayish shell, somewhat heart-shaped, with about twenty-eight flattish ribs, tranversely striated with recurved imbrications. It is a wholesome and pleasant food. Lobsters and crabs lie in wait for an opportunity of thrusting in a leg or a claw, when the Cockle is open, in order to prey on the included animal: but it often happens that the younger ones of those crustaceous animals, not being sufficiently hard to withstand the violent snapping of the shells of the larger species when they close, are deprived of the limb.

The SCALLOP is particularly remarkble for its method of moving forward upon land, or swimming upon the surface of the water. When this animal finds itself deserted by the tide, it makes very remarkable efforts to regain the water, moving towards the sea in a most singular manner. It first gapes with its shell as widely as it can, the edges being often an inch asunder; then it shuts them with a jerk, and by this the whole animal rises five or six inches from the ground. It thus tumbles any way forward, and then renews the operation until it has attained its journey's end. When in the water it is capable of supporting itself upon the surface; and there opening and shutting its shells, it tumbles over and over, and makes its way with some celerity.

The PIVOT, or RAZOR SHELL, has a very different kind of motion. As the former moves laboriously and slowly forward, so the Razor Shell has only a power of sinking point downward. The shells of this animal resemble no-

thing so much as the haft of a razor; and by this form it is
enabled to dive into the soft sand at the bottom. All the
motions of this little animal are confined to sinking or rising
a foot downwards or upwards in the sand, for it never
leaves the spot where it was first planted. From time to
time it is seen to rise about half way out of its hole; but if
any way disturbed, it sinks perpendicularly down again.
Just over the place where the Razor buries itself, there is a
small hole like a chimney, through which the animal
breathes, or imbibes the sea water. Upon the desertion of
the tide, these holes are easily distinguished by the fishermen who seek for it; and their method of enticing the
Razor up from the depth of its retreat, is, by sprinkling a
little sea-salt upon the hole. This melting no sooner
reaches the Razor below, than it rises instantly straight
upwards, and shows about half its length above the surface. This appearance, however, is instantaneous; and,
if the fisher does not seize the opportunity, the Razor
buries itself, with great ease, to its former depth. There
it continues secure; no salt can allure it a second time;
but it remains unmolested, unless the fisher will be at the
trouble of digging it out sometimes two feet below the
surface.

Such are the minute differences between bivalved shell
fish; but in the great outlines of their nature, they exactly resemble each other. It is particularly in this class of
shell fish that pearls are found in greatest abundance.
The pearl seems bred from no disorder in the animal, but
accidentally produced by the same matter that goes to form
the shell. This substance, which is soft at first, quickly
hardens; and thus, by successive coats, layer over layer,
the pearl acquires its dimensions. If cut through, it will
be found to consist of several coats, like an onion; and
sometimes a small speck is seen in the middle, upon
which the coats were originally formed.

All oysters, and most shell fish, are found to contain pearls; but that which particularly obtains the name of the pearl oyster, has a large strong whitish shell, wrinkled and rough without, and within smooth, and of a silver colour. From these the mother-of-pearl is taken, which is nothing more than the internal coats of the shell, resembling the pearl in colour and consistence. There are a great number of pearl fisheries in America and Asia. The chief of these is carried on in the Persian Gulf and at Ceylon.

The wretched people that are destined to fish for pearls, are either Negroes, or some of the poorest of the natives of Persia. The divers are not only subject to the dangers of the deep, to tempests, to suffocation at the bottom, to being devoured by sharks, but from their profession universally labour under a spitting of blood, occasioned by the pressure of air upon their lungs in going down to the bottom. The most robust and healthy young men are chosen for this employment, but they seldom survive it above five or six years. Their fibres become rigid; their eyeballs turn red; and they usually die consumptive.

It is amazing how very long they are seen to continue at the bottom. Some, as we are assured, have been known to continue three quarters of an hour under water without breathing; and to one unused to diving, ten minutes would suffocate the strongest. They fish for pearls, or rather the oysters that contain them, in boats twenty-eight feet long; and of these there are sometimes three or four hundred at a time: with each seven or eight stones, which serve for anchors. There are from five to eight divers belonging to each, that dive one after another. They are quite naked, except that they have a net hanging down from the neck to put their oysters in, and gloves on their hands to defend them while they pick the oysters from the holes in the rocks; for in this manner alone can they be gathered. Every diver is sunk by means of a stone, weighing fifty pounds, tied to the rope by which he descends. He places his foot in a kind of stirrup, and laying hold of the rope with his left hand, with his right he stops his nose to keep in his breath, as upon going down he takes in a very long inspiration. They are no sooner come to the bottom, but they give the signal to those who are in the boat to draw up the stone; which done, they go to work, filling their net as fast as they can; and then giving another signal, the boats above pull up the net loaded with oysters, and shortly after the diver himself, to take a new inspiration. They dive to the depth of fifteen fathoms, and seldom go deeper. They generally go every morning by break of day to this fatiguing employment, taking the land-wind to waft them out to sea, and returning with the sea-breeze at night. The owners of the boats usually hire the divers, and the rest of the boat's crew, as we do our labourers, at so much a day. All the oysters are brought on shore, where they are laid in a great heap till the pearl fishery is over, which continues during the months of November and December. When opportunity serves, they examine every oyster; and it is accidental whether the capture turns out advantageous.

MULTIVALVE SHELL FISH

MAY be considered as animals shut up in round boxes. Of these there are principally two kinds; such as move, and such as are stationary: the first are usually known in our cabinets by the name of sea eggs; the others are as often admired for the cavities which they scoop out for their habitation in the hardest marble. The first are called, by naturalists, Echini, or Urchins: the latter are called Pholades, or File Fish.

On a slight view, the SEA URCHIN may be compared to the husk of a chestnut; being, like it, round, and with a number of bony prickles standing out on every side. The mouth is placed downwards; the vent is above; the shell is a hollow vase, resembling a scooped apple; and this filled with a soft muscular substance, through which the intestines wind from the bottom to the top. The mouth which is placed undermost, is large and red, furnished with five sharp teeth, which are easily discerned. The jaws are strengthened by five small bones, in the centre of which is a small fleshy tongue; and from this the intestines make a winding of five spires, round the internal sides of the shell, ending at top, where the excrements are excluded. But what makes the most extraordinary part of this animal's conformation, are its horns and its spines, that point from every part of the body, like the horns of a snail, and that serve at once as legs to move upon, as arms to feel with, and as instruments of capture and defence. Between these horns it has also spines, that are not endued with such a share of motion. The spines and the horns issue from every part of its body; the spines being hard and prickly; the horns being soft, longer than the spines, and never seen except in the water. They are put forward, and withdrawn, like the horns of a snail, and are hid at the basis of the spines, serving, as was said before, for procuring food and motion. All this apparatus, however, is only seen when the animal is hunting its prey at the bottom of the water; for a few minutes after it is taken, all the horns are withdrawn into the body, and most of the spines drop off.

It is generally said of insects, that those which have the greatest number of legs, always move the slowest: but this animal seems to be an exception to the rule; for though furnished with two thousand spines, and twelve hundred horns, all serving for legs, and from their number seeming to impede each other's motion, yet it runs with some share of swiftness at the bottom, and it is sometimes no easy matter to overtake it. It is often taken upon the ebb, by following it in shallow water, either in an osier basket, or simply with the hand. Both the spines and the horns assist its motion; and the animal is usually seen running with the mouth downward.

Some kinds of this animal are as good eating as the lobster; and its eggs, which are of a deep red, are considered as a very great delicacy. But of others the taste is but indifferent; and, in all places, except the Mediterranean, they are little sought for, except as objects of curiosity. Most of this species of fish have a great variety of beautiful tints and curious forms, and many of them are highly valued in collections. Oppian tells us that the Sea Urchin was believed to have the power of uniting its dissected parts.

The Sea Urchin is oviparous, and spawns in the spring. It lives chiefly on marine worms, crabs, and other testaceous animals.

Very different in motion, though not much different in shape from these, are the ACORN shell fish, the THUMB-FOOTED shell fish, and the IMAGINARY BARNACLE. These are fixed to one spot, and appear to vegetate from a stalk. Indeed, to an inattentive spectator, each actually seems to be a kind of fungus that grows in the deep, destitute of animal life as well as motion. But the inquirer will soon change his opinion, when he comes to observe this mushroom-like figure more minutely. He will then see that the animal residing within the shell has not only life, but some degree of voraciousness. They are seen adhering to every substance that is to be met with in the ocean; rocks, roots of trees, ships' bottoms, whales, lobsters, and even crabs; like bunches of grapes clung to each other. It is amusing enough to behold their operations.* They for some time remain motionless within their shell; but when the sea is calm, they are seen opening the lid, and peeping about them; they then thrust out their long· neck, look round them for some time, and then abruptly retreat back into their box, shut their lid, and lurk in darkness and security. Some people eat them; but they are in no great repute at the tables of the luxurious, where their deformed figure would be no objection to their being introduced.

Of all animals of the shelly tribe, the PHOLADES are the most wonderful. These animals are found in different places; sometimes clothed in their proper shell, at the bottom of the water; sometimes concealed in lumps of marly earth; and sometimes lodged, shell and all, in the body of the hardest marble. In their proper shell they assume different figures; but, in general, they somewhat resemble a muscle, except that their shell is found actually composed of five or more pieces, the smaller valves serving to close up the openings left by the irregular meeting of the two principal shells. But their penetration into rocks, and their residence there, makes up the most wonderful part of their history.

This animal, when divested of its shell, resembles a roundish soft pudding, with no instrument that seems in the least fitted for boring into stones, or even penetrating the softest substance. It is furnished with two teeth indeed; but these are placed in such a situation, as to be incapable of touching the hollow surface of its stony dwelling: it

* Anderson's History of Greenland.

has also two covers to its shell, that open and shut at either end; but these are totally unserviceable to it as a miner. The instrument with which it performs all its operations, and buries itself in the hardest rocks, is only a broad fleshy substance, somewhat resembling a tongue, that is seen issuing from the bottom of its shell. With this soft, yielding instrument, it perforates the most solid marbles; and having, while yet little and young, made its way, by a very narrow entrance into the substance of the stone, it then begins to grow bigger, and thus to enlarge its apartment.

When it has buried its body in a stone, it there continues for life at its ease; the sea-water that enters at the little aperture supplying it with luxurious plenty. When the animal has taken too great a quantity of water, it is seen to spurt it out of its hole with some violence. Upon this seemingly thin diet, it quickly grows larger, and soon finds itself under a necessity of enlarging its habitation and its shell. The motion of the Pholas is slow beyond conception; its progress keeps pace with the growth of its body; and in proportion as it becomes larger, it makes its way farther into the rock. When it has got a certain way in, it then turns from its former direction, and hollows downward; till at last, when its habitation is completed, the whole apartment resembles the bowl of a tobacco pipe; the hole in the shank being that by which the animal entered.

But they are not supplied only with their rocky habitation; they have also a shell to protect them; this shell grows upon them in the body of the rock, and seems a very unnecessary addition to that defence which they have procured themselves by art. These shells take different forms, and are often composed of different number of valves; sometimes six; sometimes but three; sometimes the shell resembles a tube with holes at either end, one for the mouth, and the other for voiding the excrements.

This animal is found in greatest numbers at Ancona, in Italy; it is found along the shores of Normandy and Poitou, in France: it is found also upon some of the coasts of Scotland; and, in general, is considered as a very great delicacy at the tables of the luxurious.

CHAP. X.

Of Reptiles...The FROG... *The Toad...Varieties...Surinam Toad...Of* LIZARDS...*The Crocodile and Alligator....The Open-bellied Crocodile..The Salamander...The Cordyle, &c. ...The Iguana...The Cameleon...The Dragon...The Siren... The Tarantula...The Chalcidian Lizard.*

IF we emerge from the deep, the first and most obvious class of amphibious animals that occur upon land are Frogs and Toads.

To describe the form of animals so well known would be superfluous; to mark those differences that distinguish them from each other, may be necessary. The Frog moves by leaping; the Toad crawls along the ground: the Frog is in general less than the Toad; its colour is brighter, and with a more polished surface: the Toad is brown, rough, and dusty. The Frog is light and active, and its belly comparatively small; the Toad is slow, swollen, and incapable of escaping. The Frog, when taken, contracts itself so as to have a lump on its back; the Toad's back is straight and even. Their habitudes and manners exhibit a greater variety, and require a separate description.

The external figure of the FROG is too well known to need a description. Its power of taking large leaps is remarkably great, compared to the bulk of its body: and it is the best swimmer of all four-footed animals.

If we examine this animal internally, we shall find that it has very little brain for its size; a very wide swallow; a stomach seemingly small, but capable of great distention.

. The heart in the Frog, as in all other animals that are truly amphibious, has but one ventricle; so that the blood can circulate without the assistance of the lungs, while it keeps under water. The lungs resemble a number of small bladders joined together, like the cells of a honey-comb: they are connected to the back by muscles, and can be distended or exhausted at the animal's pleasure. Neither male nor female have any of the external instruments of generation; the anus serving for that purpose in both. Such are the most striking peculiarities in the anatomy of a Frog; and in these it agrees with the toad, the lizard, and the serpent.

The female is impregnated neither by the mouth, as some philosophers imagine, nor by the excrescence at the thumbs, as was the opinion of Linnæus; but by the inspersion of the male seminal fluid upon the eggs as they proceed from the body.

A single female produces from six to eleven hundred eggs at a time; and, in general, she throws them all out together by a single effort; though sometimes she is an hour in performing this task. It is generally in March that she deposits the ova, or spawn.

When the spawn is emitted and impregnated by the male, it drops to the bottom. The eggs, which during the four first hours suffer no perceptible change, begin then to enlarge and grow lighter; by which means they mount to the surface of the water. The twenty-first day the egg is seen to open a little on one side, and the beginning of a tail to peep out, which becomes more and more distinct every day. The thirty-ninth day the little animal begins to have motion; it moves at intervals its tail; and it is perceived that the liquor in which it is circumfused, serves it for nourishment. In two days more, some of these little creatures fall to the bottom; while others remain swimming in the fluid round them, while their vivacity and motion is seen to increase. Those which fall to the bottom remain there the whole day; but having lengthened themselves a little, for hitherto they are doubled up, they mount at intervals to the mucus, which they had quitted, and are seen to feed upon it with great vivacity. The next day they acquire their tadpole form. In three days more they are perceived to have two little fringes, that serve as fins, beneath the head; and these in four days after assume a more perfect form. It is then also that they are seen to feed very greedily upon the pond-weed. When ninety-two days old, two small feet are seen beginning to burgeon near the tail; and the head appears to be separate from the body. In five days after this, they refuse all vegetable food; their mouth appears furnished with teeth; and their hinder legs are completely formed. In this state it continues for about six or eight hours; and then the tail dropping off by degrees, the animal appears in its most perfect form.

Thus the Frog, in less than a day, having changed its figure, is seen to change its appetites also. As soon as the animal acquires its perfect state, from having fed upon vegetables it becomes carnivorous, and lives entirely upon worms and insects. But, as the water cannot supply these, it is obliged to quit its native element, and seek for food upon land, where it lives by hunting worms and taking insects by surprise.

The Frog lives for the most part out of the water; but when the cold nights begin to set in, it returns to its native element, always choosing stagnant waters, where it can lie without danger, concealed at the bottom. In this manner it continues torpid, or with but very little motion, all the winter; like the rest of the dormant race, it requires no food: and the circulation is slowly carried on without any assistance from the air. In the countries round Hudson's Bay, it is often found frozen hard, in which state it is as brittle as glass; yet by wrapping it in warm skins, and exposing it to a slow fire, it will return to life.

The difference of sexes, which was mentioned above, is not perceivable in these animals, until they have arrived at their fourth year; nor do they begin to propagate, till they have completed that period. By comparing their slow growth with their other habitudes, it would appear, that they live about twelve years; but having so many enemies, both by land and water, it is probable that few of them arrive at the end of their term.

Frogs live upon insects of all kinds; but they never eat any, unless they have motion. They continue fixed and immoveable till their prey appears; and just when it comes sufficiently near, they jump forward with great agility, dart out their tongues, and seize it with certainty. The tongue in this animal, as in the toad, lizard, and serpent, is extremely long, and formed in such a manner that it swallows the point down its throat; so that a length of tongue is thus drawn out, like a sword from its scabbard, to assail its prey. This tongue is furnished with with a glutinous substance; and whatever insect it touches infallibly adheres, and is thus held fast till it is drawn into the mouth.

The croaking of Frogs is well known; whence in some countries they are distinguished by the ludicrous title of Dutch Nightingales. The Large Water or Bull Frogs of the northern countries have a note as loud as the bellowing of a bull; and, for this purpose, puff up the cheeks to a surprising magnitude. Of all Frogs, however, the male only croaks; the female is silent; before wet weather their voices are in full exertion; they are then heard with unceasing assiduity, sending forth their call, and welcoming the approaches of their favourite moisture. No weather-glass was ever so true as a frog, in foretelling an approaching change. This may probably serve to explain an opinion which some entertain, that there is a month in the year, called Paddock Moon, in which the Frogs never croak: the whole seems to be no more than that, in the hot season, when the moisture is dried away, and consequently, when these animals neither enjoy the quantity of health nor food that at other times they are supplied with, they show by their silence how much they are displeased with the weather.

As Frogs adhere closely to the backs of their own species, so it has been found, by repeated experience, they will also adhere to the backs of fishes. Few that have ponds, but know that these animals will stick to the backs of carp, and fix their fingers in the corner of each eye. In this manner they are often caught together; the carp blinded, and wasted away.

The EDIBLE FROG, which is considerably larger than the common species, is rare in England, but is abundant in Italy, France, and Germany, where its hind quarters are looked upon as a delicacy. It is of an olive green hue, marked with black patches on its back, and on its limbs with black transverse bars.

The TREE FROG is small, slender, and elegantly made; green in all the upper parts, whitish in the abdomen, and reddish on the under surface of the limbs. In summer it resides principally on the upper branches of trees, where it feeds on insects, which it catches very dexterously. It is remarkable for its power of absorbing water. It is found in France, Germany, and Italy, and other European countries, and in various parts of America; but not in Great Britain.

THE TOAD.

IF we regard the figure of the Toad, there seems nothing in it that should disgust, more than that of the frog. Its form and proportions are nearly the same; and it chiefly differs in colour, which is blacker; and its slow and heavy motion, which exhibits nothing of the agility of the frog: yet such is the force of habit, begun in early prejudice, that those who consider the one as a harmless, playful animal, turn from the other with horror and disgust. The frog is considered as a useful assistant in ridding our grounds of vermin; the Toad as a secret enemy, that only wants an opportunity to infect us with its venom.

As the Toad bears a general resemblance in figure to the frog, so also it resembles that animal in its nature and appetites. When, like the frog, these animals have undergone all the variations of their tadpole state, they forsake

the water, and are often seen, in a moist summer's evening, crawling up, by myriads, from fenny places, into drier
situations.   There, having found out a retreat, or having
dug themselves one with their mouth and hands, they lead
a patient solitary life, seldom venturing out, except when
the moisture of a summer's evening invites them abroad.
At that time the grass is filled with snails, and the path-
ways covered with worms, which make their principal food.
Insects also, of every kind, they are fond of; and we have
the authority of Linnæus for it, that they sometimes continue
immoveable, with the mouth open, at the bottom of shrubs,
where the butterflies, in some measure fascinated, are seen
to fly down their throats.

In a letter from Mr. Arscott, there are some curious
particulars relating to this animal, which throw great light
upon its history.   "Concerning the Toad," says he, "that
lived so many years with us, and was so great a favourite,
the greatest curiosity was its becoming so remarkably
tame: it had frequented some steps before our hall door
some years before my acquaintance commenced with it,
and had been admired by my father for its size (being the
largest I ever met with) who constantly paid it a visit every
evening.   I knew it myself above thirty years; and by
constantly feeding it, brought it to be so tame, that it always came to the candle and looked up, as if expecting to
be taken up and brought upon the table, where I always
fed it with insects of all sorts.   It was fondest of flesh
maggots, which I kept in bran; it would follow them, and
when within a proper distance, would fix his eyes, and remain motionless, for near a quarter of a minute, as if preparing for the stroke, which was an instantaneous throwing of its tongue at a great distance upon the insect, which
stuck to the tip by a glutinous matter.   The motion is
quicker than the eye can follow.   I cannot say how long

my father had been acquainted with the Toad, before I knew it; but when I was first acquainted with it, he used to mention it as the old Toad I have known so many years. I can answer for thirty-six years. This old Toad made its appearance as soon as the warm weather came; and I always concluded it retired to some dry bank, to repose till spring. When we new laid the steps, I had two holes made in the third step on each side, with a hollow of more than a yard long for it; in which I imagine it slept, as it came thence at its first appearance. It was seldom provoked. Neither that Toad, nor the multitudes I have seen tormented with great cruelty, ever showed the least desire of revenge, by spitting or emitting any juice from 'their pimples. Sometimes, upon taking it up, it would let out a great quantity of clear water, which, as I have often seen it to do the same upon the steps when quite quiet, was certainly its urine, and no more than a natural evacuation Spiders, millepedes, and flesh maggots, seem to be this animal's favourite food. I imagine if a bee were to be put before a Toad, it would certainly eat it to its cost;* but as bees are seldom stirring at the same time that Toads are, they rarely come in their way: as they do not appear after sun-rising, or before sun-set. In the heat of the day they will come to the mouth of their hole, I believe for air. I once, from my parlour window, observed a large Toad I had in the bank of a bowling-green, about twelve at noon, on a very hot day, very busy and active upon the grass. So uncommon an appearance made me go out to see what it was; when I found an innumerable swarm of winged ants had dropped round his hole; which temptation was as irresistible as a turtle would be to a luxurious alder-

---

* Ræsel tried a frog; it swallowed the bee alive; its stomach was stung, and the animal vomited it up again.

man. In respect to its end, had it not been for a tame raven, I make no doubt but it would have been now living. This bird, one day seeing it at the mouth of its hole, pulled it out; and, although I rescued it, pulled out one eye, and hurt it so, that, notwithstanding its living a twelvemonth, it never enjoyed itself, and had a difficulty of taking its food, missing the mark for want of its eye. Before that accident, it had all the appearance of perfect health."

The Toad, contrary to vulgar prejudice, is a harmless, defenceless creature, torpid and unvenomous, and seeking the darkest retreats, not from the malignity of its nature, but the multitude of its enemies.

Like all of the Frog kind, the Toad is torpid in winter. It chooses then for a retreat either the hollow root of a tree, the cleft of a rock, or sometimes the bottom of a pond, where it is found in a state of seeming insensibility. As it is very long-lived, it is very difficult to be killed; its skin is tough, and cannot be easily pierced; and, though covered with wounds, the animal continues to show signs of life, and every part appears in motion. But what shall we say to its living for centuries lodged in the bosom of a rock, or cased within the body of an oak tree, without the smallest access on any side, either for nourishment or air, and yet taken out alive and perfect! Stories of this kind it would be as rash to contradict as it is difficult to believe; we have the highest authorities bearing witness to their truth, and yet the whole analogy of nature seems to arraign them of falsehood. Bacon asserts, that Toads are found in this manner; Dr. Plot asserts the same: there is, to this day, a marble chimney-piece at Chatsworth, with the print of the Toad upon it, and tradition of the manner in which it was found. In the Memoirs of the Academy of Sciences, there is an account of a Toad found alive and healthy in the heart of a very thick elm, without the smallest entrance or egress.* In the year 1731, there was another found near Nantz, in the heart of an old oak, without the smallest issue to its cell; and the discoverer was of opinion, from the size of the tree, that the animal could not have been confined there less than eighty or a hundred years, without sustenance and without air.

Of this animal there are several varieties; such as the water and the land Toad, which probably differ only in the ground colour of their skin. In the first it is more inclining to ash colour, with brown spots; in the other the colour is brown, approaching to black. The water Toad is not so large as the other; but both equally breed in that element. The size of the Toad with us is generally from two to four inches long; but in the fenny countries of Europe they are seen much larger; and not less than a common crab. But this is nothing to what they are found in some of the tropical climates, where travellers often, for the first time, mistake a Toad for a tortoise. Their usual size is from six to seven inches; but there are some still larger, and as broad as a plate. Of these, some are beautifully streaked and coloured; some studded over, as if with pearls; others bristled with horns or spines; some have the head distinct from the body, while others have it so sunk in that the animal appears without a head. With us the opinion of its raining Toads and Frogs has long been justly exploded; but it still is entertained in the tropical countries, and that not only by the savage natives, but the more refined settlers; who are apt enough to add the prejudices of other nations to their own.

It would be a tedious as well as a useless task to enter into all the minute discriminations of these animals, as found in different countries or places; but the PIPA, or the

* Vide the year 1719.

Surinam Toad, is too strange a creature not to require an exact description.

This animal is in form more hideous than even the common Toad. The body is flat and broad; the head small; the jaws, like those of the mole, are extended, and evidently formed for rooting in the ground; the skin of the neck forms a sort of wrinkled collar; the colour of the head is of a dark chestnut, and the eyes are small; the back, which is very broad, is of a lightish gray, and seems covered over with a number of small eyes, which are round, and placed at nearly equal distances. These eyes are very different from what they seem; they are the animal's eggs covered with their shells, and placed there for hatching. They are generated within the female, who drops them on the ground. The male then collects them, and deposits them carefully on the back of the female, where, after impregnation, they are pressed into the cellules, which close upon them. These eggs are buried deep in the skin, and in the beginning of gestation but just appear; and are very visible when the young animal is about to burst from its confinement. They are of a reddish shining yellow colour; and the spaces between them are full of small warts resembling pearls.

In this manner the Pipa is seen travelling with her wondrous family on her back, in all the different stages of maturity. Some of the strange progeny not yet come to sufficient perfection, appear quite torpid, and as yet without life in the egg: others seem just beginning to rise through the skin; here peeping forth from the shell, and there having entirely forsaken their prison: some are sporting at large upon their parent's back; and others descending to the ground, to try their own fortune below.

OF LIZARDS.

IT is no easy matter to tell to what class in nature Lizards are chiefly allied. They are unjustly raised to the rank of beasts, as they bring forth eggs, dispense with breathing, and are not covered with hair. They cannot be placed among fishes, as the majority of them live upon the land: they are excluded from the serpent tribe, by their feet, upon which they run with some celerity; and from the insects, by their size; for though the newt may be looked upon in this contemptible light, a crocodile would be a terrible insect indeed.

As Lizards thus differ from every other class of animals, they also differ widely from each other. With respect to size, no class of beings has its ranks so opposite.

The colour of these animals also is very various; as they are found of a hundred different hues: green, blue, red, chesnut, yellow, spotted, streaked, and marbled. Were colour alone capable of constituting beauty, the Lizard would often please; but there is something so repulsive in the animal's figure, that the brilliancy of its scales, or the variety of its spots, only tend to give an air of more exquisite venom, of greater malignity. The figure of these animals is not less various; sometimes swollen in the belly, sometimes pursed up at the throat; sometimes with a rough set of spines on the back, like the teeth of a saw; sometimes with teeth, at others with none; sometimes venomous, at others harmless, and even philanthropic; sometimes smooth and even; sometimes with a long, slender tail; and often with a shorter blunt one.

But their greatest distinction arises from their manner of bringing forth their young: some of them are viviparous; some are oviparous; and some bring forth small spawn, like fishes.

The only animals of this genus which are common in Great Britain are, the SCALY LIZARD, which is about six inches in length; the BROWN LIZARD, or EFT, which is about three inches long; and the WARTY LIZARD, or SALAMANDER, of which we shall presently treat more at large.

THE CROCODILE

Is an animal placed at a happy distance from the inhabitants of Europe, and formidable only in those regions where men are scarce, and arts are little known. To look for it in all its terrors, we must go to the uninhabited regions of Africa, Asia, and America. It is, however, a native of Egypt, where it was formerly worshipped.

Of this terrible animal there are two kinds; the Crocodile, properly so called, and the Cayman or Alligator. Travellers, however, have rather made the distinction than Nature; for in the general outline, and in the nature of these two animals, they are entirely the same. The distinctions usually made between the Crocodile and Alligator are these: the body of the Alligator is less slender than that of the Crocodile; and its snout is broad, blunt, and less produced than that of the true Crocodile. The fourth tooth on each side of the lower jaw of the Alligator enters a hole in the upper when the mouth is closed, and the toes are only half webbed.

This animal grows to a great length, being sometimes found thirty feet long from the tip of the snout to the end of the tail; its most usual length, however, is eighteen. One which was dissected by the Jesuits at Siam was eighteen feet and a half, French measure, in length; of which the tail was no less than five feet and a half, and the head and neck above two feet and a half. It was four feet nine inches in circumference, where thickest. The fore legs had the same parts and conformation as the arms of a man both within and without. The hands, if they may be so called, had five fingers; the two last of which had no nails, and were of a conical figure. The hinder legs, including the thigh and paw, were two feet two inches long; the paws, from the joint to the extremity of the longest claws, were about nine inches; they were divided into four toes, of which three were armed with large claws, the longest of which was an inch and a half; these toes were united by a membrane, like those of a duck, but much thicker. The head was long, and had a little rising at the top; but the rest was flat, and especially towards the extremity of the jaws. It was covered by a skin, which adhered firmly to the skull and to the jaws. The skull was rough and unequal in several places. The eye was very small in proportion to the rest of the body. The jaws seemed to shut one upon the other; and nothing can be more false than that the animal's under jaw is without motion; it moves,

like the lower jaw in all other animals, while the upper is
fixed to the skull, and absolutely immoveable. The animal had twenty-seven cutting teeth in the upper jaw, and
fifteen in the lower, with several void spaces between
them. The distance of the two jaws, when opened as wide
as they could be, was fifteen inches and a half; this is a
very wide yawn, and could easily enough take in the body
of a man. From the shoulders to the extremity of the tail,
the animal was covered with large scales, of a square form,
disposed like parallel girdles. The creature was covered
not only with these, but all over with a coat of armour;
which, however, contrary to what has been commonly asserted, was not proof against a musket-ball. It had no
bladder; but the kidney sent the urine to be discharged by
the anus. There were sixty-two joints in the backbone,
which, though very closely united, had sufficient play to enable the animal to bend like a bow to the right and the left;
so that what we hear of escaping the creature by turning
out of the right line, and of the animal's not being able to
wheel readily after its prey, seems to be fabulous.

Such is the figure and conformation of this formidable
animal, that depopulates countries, and makes the most
navigable rivers desert and dangerous. They are seen in
some places lying for whole hours, and even days, stretched
in the sun, and motionless; so that one not used to them
might mistake them for trunks of trees, covered with a
rough and dry bark; but the mistake would soon be fatal, if
not prevented: for the torpid animal, at the near approach
of any living thing, darts upon it with instant swiftness, and
at once drags it down to the bottom. In the times of an
inundation, they sometimes enter the cottages of the natives, where the dreadful visitant seizes the first animal it
meets with. There have been several examples of their
taking a man out of a canoe in the sight of his companions,
without their being able to lend him any assistance.

The strength of every part of the Crocodile is very great; and its arms, both offensive and defensive, irresistible. Most naturalists have remarked, from the shortness of its legs, the amazing strength of the tortoise : but what is the strength of such an animal, compared to that of the Crocodile, whose legs are very short, and whose size is so superior? Its principal instrument of destruction is, the tail ; with a single blow of this it has often overturned a canoe, and seized upon the poor savage, its conductor.

Though not so powerful, yet it is very terrible even upon land. The Crocodile seldom, except when pressed by hunger, or with a view of depositing its eggs, leaves the water. Its usual method is to float along upon the surface, and seize whatever animals come within its reach ; but when this method fails, it then goes closer to the bank. Disappointed of its fishy prey, it there waits, covered up among the sedges, in patient expectation of some land animal that may come to drink ; the dog, the bull, the tiger, or man himself. Nothing is to be seen of the insidious destroyer as the animal approaches ; nor is its retreat discovered till it be too late for safety. It seizes the victim with a spring, and goes at a bound much faster than so unwieldy an animal could be thought capable of ; then, having secured the creature with both teeth and claws, it drags it into the water, instantly sinks with it to the bottom, and in this manner quickly drowns it.

Sometimes it happens that the creature the Crocodile has thus surprised escapes from its grasp wounded, and makes off from the river-side. In such a case, the tyrant pursues with all its force, and often seizes it a second time : for, though seemingly heavy, the Crocodile runs with great celerity. In this manner it is sometimes seen above half a mile from the bank, in pursuit of an animal wounded beyond the power of escaping, and then dragging it back to the river-side, where it feasts in security.

It often happens, in its depredations along the bank, that the Crocodile seizes on a creature as formidable as itself, and meets with the most desperate resistance. We are told of frequent combats between the Crocodile and the tiger. All creatures of the tiger kind are continually oppressed by a parching thirst, that keeps them in the vicinity of great rivers, whither they descend to drink very frequently. It is upon these occasions that they are seized by the Crocodile; and they die not unrevenged. The instant they are seized upon, they turn with the greatest agility, and force their claws into the Crocodile's eyes, while he plunges with his fierce antagonist into the river. There they continue to struggle for some time, till at last the tiger is drowned.

In this manner the Crocodile seizes and destroys all animals, and is equally dreaded by all. There is no animal, but man alone, that can combat it with success. We are assured by Labat, that a negro, with no other weapons than a knife in his hand, and his left arm wrapped round with a cow's hide, ventures boldly to attack the animal in his own element. As soon as he approaches the Crocodile, he presents his left arm, which the animal swallows most greedily; but sticking in his throat, the negro has time to give it several stabs under the throat; and the water also getting in at the mouth, which is held involuntarily open, the creature is soon bloated up as big as a tun, and expires.

Whatever be the truth of these accounts, certain it is that Crocodiles are taken by the Siamese in great abundance. The manner of taking them is, by throwing three or four strong nets across a river, at proper distances from each other; so that, if the animal breaks through the first, it may be caught by one of the rest. When it is first taken, it employs its tail with great force; but, after many unsuccessful struggles, it is at last exhausted. Then the natives approach their prisoner in boats, and pierce him with their weapons in the most tender parts, till he is weakened with the loss of blood. When he has done stirring, they begin by tying up his mouth, and, with the same cord, they fasten his head to his tail; which last they bend back like a bow. They are not, however, yet perfectly secure from his fury; but, for their greater safety, they tie his fore feet, as well as those behind, to the top of his back.

The Crocodile, thus brought into subjection, or bred up young, is used to divert and entertain the great men of the East. It is often managed like a horse: a curb is put into its mouth, and the rider directs it as he thinks proper. Though awkwardly formed, it does not fail to proceed with some degree of swiftness, and is thought to move as fast as some of the most unwieldy of our own animals, the hog or the cow.

Along the rivers of Africa this animal is sometimes taken in the same manner as the shark. Several Europeans go together in a large boat, and throw out a piece of beef upon a hook and strong fortified line, which the Crocodile seizing and swallowing, is drawn along, floundering and struggling, until its strength is quite exhausted, when it is pierced in the belly, which is its tenderest part; and thus, after numberless wounds, is drawn ashore. In this part of the world also, as well as at Siam, the Crocodile makes an object of savage pomp, near the palaces of their monarchs. Phillips informs us, that at Sabi, on the Slave Coast, there are two pools of water near the royal palace, where Crocodiles are bred, as they breed carp in the ponds in Europe.

There is a very powerful smell of musk about all these animals. Travellers are not agreed in what part of the body these musk-bags are contained; but the most probable opinion is, that this substance is amassed in glands under the legs and arms. The Crocodile's flesh is, at best, very bad, tough eating; but, unless the musk-bags be separated, it is insupportable. The Negroes themselves cannot well digest the flesh; but a Crocodile's egg is to them the most delicate morsel in the world.

All Crocodiles breed near fresh waters; and for this purpose the female, when she comes to lay, chooses a place by the side of a river or some fresh water lake, to deposit her brood in. She always pitches upon an extensive sandy shore, where she may dig a hole without danger of detection from the ground being fresh turned up. There she deposits from eighty to a hundred eggs, of the size of a tennis-ball, and of the same figure, covered with a tough, white skin, like parchment. She takes above an hour to perform this task; and then covering up the place so artfully that it can scarcely be perceived, she goes back, to return again the next day. Upon her return, with the same precaution as before, she lays about the same number of eggs; and the day following also a like number. Thus, having deposited her whole quantity, and having covered them close up in the sand, they are soon vivified by the heat of the sun; and at the end of thirty days the young ones begin to break open the shell. At this time, the female is instinctively taught that her young ones want relief; and she goes upon land, to scratch away the sand, and set them free. Her brood quickly avail themselves of their liberty; a part run unguided to the water; another part ascend the back of the female, and are carried thither in greater safety. But the moment they arrive at the water, all natural connexion is at an end. The whole brood scatters into different parts of the bottom: by far the greater number are destroyed; and the rest find safety in their agility or minuteness.

The open-bellied Crocodile is furnished with a false belly, like the opossum, where the young creep out and in, as their dangers and necessities require. It is probable that this open-bellied Crocodile is viviparous, and fosters her young, that are prematurely excluded, in this second womb, until they come to proper maturity.

THE ALLIGATOR, OR AMERICAN CROCODILE,

WHICH is called the Cayman by the Indians, is closely allied to the preceding species; the principal difference between them being that its head and part of its neck are much more smooth than those of the latter, and that its snout is more wide and flat, and more rounded at the extremity. The usual length of the Alligator is seventeen or eighteen feet, but it sometimes exceeds this. This animal is a native of the warmer parts of America, in some of which it is astonishingly numerous. Its voice is loud and dreadful; and its musky scent is sometimes so powerful as to be exceedingly offensive. M. Pagés tells us that, near an American river, which was thronged with Alligators, the effluvia was so strong as to impregnate his provisions with the nauseous effluvia of rotten musk. Bertram, in his Travels through the Southern States of North America, which were published in 1774, has given an amusing account of the observations which he made on the Alligators in the river St. Juan, in East Florida, and of the dangers to which he was exposed from these amphibious furies. He thus describes a battle between two of them: "Behold him rushing forth from the flags and reeds! His enormous body swells; his plaited tail, brandished high, floats upon the lake ; the waters, like a cataract, descend from his opening jaws ; clouds of smoke issue from his dilated nostrils ; the earth trembles with his thunder. When immediately, from the opposite coast of the Lagoon, emerges from the deep his rival champion. They suddenly dart upon each other. The boiling surface of the lake marks their rapid course, and a terrible conflict commences. They now sink to the bottom, folded together in horrid wreaths. The water becomes thick and discoloured. Again they sink, when the contest ends at the oozy bottom of the lake, and the vanquished makes a hazardous escape, hiding himself in the muddy turbulent waters and sedge on a distant shore."

During the hot months, in South America, these creatures bury themselves in the mud, and become torpid. M. de Humboldt gives an amusing instance of one of them getting up from his two months' nap. "Sleeping with one of his friends on a bench covered with leather, Don Miguel," says he, " was awakened early in the morning by violent shakes and a horrible noise. Clods of earth were thrown into the middle of the hut. Presently a young Crocodile, two or three feet long, issued from under the bed, darted at a dog that lay on the threshold of the door, and, missing him in the impetuosity of his spring, ran towards the beach to attain the river. On examining the spot where the *barbacon*, or bedstead, was placed, the cause of this strange adventure was easily discovered. The ground was disturbed to a considerable depth. It was dried mud, that had covered the Crocodile in that state of lethargy, or summer sleep, in which many of the species lie during the absence of the rains amid the Llanos. The noise of men and horses, perhaps the smell of the dog, had awakened the Crocodile. The hut being placed at the edge of the pool, and inundated during part of the year, the Crocodile had, no doubt, entered at the time of the inundations of the Savannahs, by the same opening by which M. Pozo saw it go out."

The habits of the North American Alligator are described with great accuracy, and in a very amusing manner, by Mr. Audubon. "One of the most remarkable objects connected with the Natural History of America, that attract the traveller's eye as he ascends through the mouths of the mighty sea-like river Mississippi, is the Alligator. There, along the muddy shores, and on the large floating logs, these animals are seen lying stretched at full length, basking and asleep, or crossing to and fro the stream in search of food, with the head only out of the water. It is here neither wild nor shy; nor is it the very dangerous animal represented by travellers. But, to give you details that probably may not be uninteresting, I shall describe their more private haunts, and relate what I have experienced and seen respecting them in their habits.

"In Louisiana, all our lagoons, bayous, creeks, ponds, lakes and rivers, are well stocked with them; they are found wherever there is a sufficient quantity of water to hide them, or to furnish them with food; and they continue thus, in great numbers, as high as the mouth of the Arkansas River, extending east to North Carolina, and as far west as I have penetrated. On the Red River, before it was navigated by steam vessels, they were so extremely abundant that, to see hundreds at a sight along the shores, or on the immense rafts of floating or stranded timber, was quite a common occurrence, the smaller on the backs of the larger, groaning and uttering their bellowing noise, like thousands of irritated bulls about to meet in fight, but all so careless of man that, unless shot at or positively disturbed, they remained motionless, suffering boats or canoes to pass within a few yards of them, without noticing them in the least. The shores are yet trampled by them in such a manner, that their large tracks are seen as plentiful as those of sheep in a fold. It was on that river particularly thousands of the largest size were killed, when the mania of having shoes, boots, or saddle-seats, made of their hides lasted. It had become an article of trade, and many of the squatters and strolling Indians followed for a time no other business. The discovery that their skins are not sufficiently firm and close-grained to prevent water or dampness long, put a stop to their general destruction, which had already become very apparent. The leather prepared from these skins was handsome and very pliant, exhibiting all the regular lozenges of the scales, and able to receive the highest degree of polish and finishing.

"The usual motion of the Alligator, when on land, is slow and sluggish; it is a kind of laboured crawling, performed by moving alternately each leg, in the manner of a quadruped when walking, scarce able to keep up their weighty bodies from dragging on the earth, and leaving the track of their long tail on the mud, as if that of the keel of a small vessel. Thus they emerge from the water, and go about the shores and the woods, or the fields, in search of food, or of a different place of abode, or one of safety to deposit their eggs. If, at such times, when at all distant from the water, an enemy is perceived by them, they droop and lie flat, with the nose on the ground, watching the intruder's movements with their eyes, which are able to move considerably round, without affecting the position of the head. Should a man then approach them, they do not attempt either to make away or attack, but merely raise their body from the ground for an instant, swelling themselves, and issuing a dull blowing, not unlike that of a blacksmith's bellows. Not the least danger need be apprehended: then you either kill them with ease, or leave them. But, to give you a better idea of the slowness of their movements and progress of travels on land, when arrived at a large size, say twelve or fifteen feet, believe me when I tell you, that having found one in the morning, fifty yards from a lake, going to another in sight, I have left him unmolested, hunted through the surrounding swamps all the day, and met the same Alligator within five hundred yards of the spot when returning to my camp at dusk. On this account they usually travel during the night, they being then less likely to be disturbed, and having a better chance to surprise a litter of pigs, or of land tortoises, for prey.

" The power of the Alligator is in his great strength; and the chief means of his attack or defence is his large tail, so well contrived by nature to supply his wants, or guard him from danger, that it reaches, when curved into half a circle, his enormous mouth. Woe be to him who goes within the reach of this tremendous thrashing instrument; for no matter how strong or muscular—if human, he must suffer greatly, if he escapes with life. The monster, as he strikes with this, forces all objects within the circle towards his jaws, which, as the tail makes a motion, are opened to their full stretch, thrown a little sideways, to receive the object, and, like battering-rams, to bruise it shockingly in a moment.

" The Alligator, when after prey in the water, or at its edge, swims so slowly towards it, as not to ruffle the water. It approaches the object sideways, body and head all concealed, till sure of his stroke; then, with a tremendous blow, as quick as thought, the object is secured, as I described before.

" When Alligators are fishing, the flapping of their tails about the water may be heard at half a mile ; but, to describe this in a more graphic way, suffer me to take you along with me, in one of my hunting excursions, accompanied by friends and negroes. In the immediate neighbourhood of Bayou-Sarah, on the Mississippi, are extensive shallow lakes and morasses ; they are yearly overflowed by the dreadful floods of that river, and supplied with myriads of fishes of many kinds, amongst which trouts are most abundant, white perch, catfish, and alligator gars, or devil fish. Thither, in the early part of autumn, when the heat of a southern sun has exhaled much of the water, the squatter, the planter, the hunter, all go in search of sport. The lakes are then about two feet deep, having a fine sandy bottom ; frequently much grass grows in them, bearing crops of seed, for which multitudes of water-fowl resort to those places. The edges of these lakes are deep swamps, muddy for some distance, overgrown with heavy large timber, principally cypress, hung with Spanish beard, and tangled with different vines, creeping plants, and cane, so as to render them almost dark during the day, and very difficult to the hunter's progress. Here and there in the lakes are small islands, with clusters of the same trees, on which flocks of snake-birds, wood-ducks, and different species of herons, build their nests. Fishing-lines, guns, and rifles, some salt, and some water, are all the hunters take. Two negroes precede them, the woods are crossed—the scampering deer is seen—the racoon and the opossum cross before you—the black, the gray, and the fox squirrel, are heard barking—here on a tree close at hand is seen an old male pursuing intensely a younger one ; he seizes it, they fight desperately, but the older attains his end, *vincit castratque juniorem.* (Now, my dear sirs, if this is not mental power illustrated, what shall we call it?) As you proceed further on, the *hunk hunk* of the lesser ibis is heard from different parts, as they rise from the puddles that supply them with crayfishes. At last the opening of the lake is seen: it has now become necessary to drag oneself along through the deep mud, making the best of the way, with the head bent, through the small brushy growth, caring about nought but the lock of your gun. The long narrow Indian canoe kept to hunt those lakes, and taken into them during the fresh, is soon launched, and the party seated in the bottom, is paddled or poled in search of water-game. There, at a sight, hundreds of Alligators are seen dispersed over all the lake, their head, and all the upper part of the body, floating like a log, and, in many instances, so resembling one that it requires to be accustomed to see them to know the distinc-

tion. Millions of the large wood-ibis are seen wading through the water, mudding it up, and striking deadly blows with their bills on the fish within. Here are a hoard of blue herons—the sand-hill crane rises with hoarse note—the snake-birds are perched here and there

on the dead timber of the trees—the cormorants are fishing—buzzards and carrion-crows exhibit a mourning train,
patiently waiting for the water to dry and leave food for
them—and far in the horizon the eagle overtakes a devot-
ed wood-duck, singled from the clouded flocks that have
been bred there. It is then that you see and hear the
Alligator at his work,—each lake has a spot deeper than
the rest, rendered so by those animals who work at it, and
always situate at the lower end of the lake, near the connecting bayous, that, as drainers, pass through all those
lakes, and discharge sometimes many miles below where
the water had made its entrance above, thereby insuring
to themselves water as long as any will remain. This is
called by the hunters the Alligators' Hole. You see them
there lying close together. The fish that are already dy-
ing by thousands, through the insufferable heat and stench
of the water, and the wounds of the different winged enemies constantly in pursuit of them, resort to the Alligators'
Hole to receive refreshment, with a hope of finding security also, and follow down the little currents flowing
through the connecting sluices: but no! for, as the water
recedes in the lake, they are here confined. The Alligators thrash them and devour them whenever they feel hungry, while the ibis destroys all that make towards the
shore. By looking attentively on this spot, you plainly
see the tails of the Alligators moving to and fro, splashing,
and now and then, when missing a fish, throwing it up in
the air. The hunter, anxious to prove the value of his
rifle, marks one of the eyes of the largest Alligator, and,
as the hair trigger is touched, the Alligator dies. Should
the ball strike one inch astray from the eye, the animal
flounces, rolls over and over, beating furiously about him
with his tail, frightening all his companions, who sink immediately, whilst the fishes, like blades of burnished metal,

leap in all directions out of the water, so terrified are they at this uproar.* Another and another receives the shot in the eye, and expires; yet those that do not feel the fatal bullet pay no attention to the death of their companions till the hunter approaches very close, when they hide themselves for a few moments by sinking backwards.

"So truly gentle are the Alligators at this season, that I have waded through such lakes in company of my friend Augustin Bourgeat, Esq. to whom I owe much information, merely holding a stick in one hand to drive them off, had they attempted to attack me. When first I saw this way of travelling through the lakes, waist-deep, sometimes with hundreds of these animals about me, I acknowledge to you that I felt great uneasiness, and thought it fool-hardiness to do so: but my friend, who is a most experienced hunter in that country, removed my fears by leading the way, and, after a few days, I thought nothing of it. If you go towards the head of the Alligator, there is no danger, and you may safely strike it with a club, four feet long, until you drive it away, merely watching the operations of the point of the tail, that, at each blow you give, thrashes to the right and left most furiously.

"The drivers of cattle from the Appelousas, and those of mules from Mexico, on reaching a lagoon or creek, send several of their party into the water, armed merely each with a club, for the purpose of driving away the Alligators from the cattle; and you may then see men, mules, and those monsters, all swimming together, the men striking the Alligators, that would otherwise attack the cattle, of

* This so alarms the remaining Alligators, that, regularly, in the course of the following night, every one, large and small, removes to another hole, going to it by water, and probably for a week not one will be seen there.

which they are very fond, and those latter hurrying towards the opposite shores, to escape those powerful enemies. They will swim swiftly after a dog, or a deer, or a horse, before attempting the destruction of man, of which I have always remarked they were afraid, if the man feared not them.

"Although I have told you how easily an Alligator may be killed with a single rifle-ball, if well aimed, that is to say, if it strike either in the eye or very immediately above it, yet they are quite as difficult to be destroyed if not shot properly ; and, to give you an idea of this, I shall mention two striking facts.

"My good friend Richard Harlan, M. D. of Philadelphia, having intimated a wish to have the heart of one of these animals to study its comparative anatomy, I one afternoon went out about half a mile from the plantation, and, seeing an Alligator that I thought I could put whole into a hogshead of spirits, I shot it immediately on the skull bone. It tumbled over from the log on which it had been basking into the water, and, with the assistance of two negroes, I had it out in a few minutes, apparently dead. A strong rope was fastened round its neck, and, in this condition, I had it dragged home across logs, thrown over fences, and handled without the least fear. Some young ladies there, anxious to see the inside of his mouth, requested that the mouth should be propped open with a stick put vertically ; this was attempted, but at this instant the first stunning effect of the wound was over, and the animal thrashed and snapped its jaws furiously, although it did not advance a foot. The rope being still round the neck, I had it thrown over a strong branch of a tree in the yard, and hauled the poor creature up swinging, free from all about it, and left it twisting itself, and scratching with its fore feet to disengage the rope. It remained in this condition until the next morning, when finding it still alive, though very weak, the hogshead of spirits was put under it, and the Alligator fairly lowered into it with a surge. It twisted about a little, but the cooper secured the cask, and it was shipped to Philadelphia, where it arrived in course.

" Again, being in company with Augustin Bourgeat, Esq., we met an extraordinary large Alligator in the woods whilst hunting; and, for the sake of destruction I may say, we alighted from our horses, and approached with full intention to kill it. The Alligator was put between us, each of us provided with a long stick to irritate it; and, by making it turn its head partly on one side, afford us the means of shooting it immediately behind the fore leg and through the heart. We both discharged five heavy loads of duck-shot into its body, and almost all into the same hole, without any other effect than that of exciting regular strokes of the tail, and snapping of the jaws at each discharge, and the flow of a great quantity of blood out of the wound, and mouth, and nostrils of the animal: but it was still full of life and vigour, and to have touched it with the hand would have been madness; but as we were anxious to measure it, and to knock off some of its larger teeth to. make powder charges, it was shot with a single ball just above the eye, when it bounded a few inches off the ground, and was dead when it reached it again. Its length was seventeen feet; it was apparently centuries old; many of its teeth measured three inches. The shot taken were without a few feet only of the circle that we knew the tail could form, and our shots went en masse.

" As the lakes become dry, and even the deeper connecting bayous empty themselves into the rivers, the Alligators congregate into the deepest hole in vast numbers; and, to this day, in such places, are shot for the sake of their oil, now used for greasing the machinery of steam-engines and cotton-mills, though formerly, when indigo was made in Louisiana, the oil was used to assuage the overflowing of the boiling juice, by throwing a ladleful into the kettle whenever this was about to take place. The Alligators are caught frequently in nets by fishermen; they then come without struggling to the shore, and are killed by blows on the head given with axes.

"When autumn has heightened the colouring of the foliage of our woods, and the air feels more rarefied during the nights and earlier part of the day, the Alligators leave the lakes to seek for winter quarters, by burrowing under the roots of trees, or covering themselves simply with earth along their edges. They become then very languid and inactive, and, at this period, to sit or ride on one would not be more difficult than for a child to mount his wooden rocking-horse. The negroes, who now kill them, put all danger aside, by separating, at one blow with an axe, the tail from the body. They are afterwards cut up in large pieces, and boiled whole in a good quantity of water, from the surface of which the fat is collected with large ladles. One single man kills oftentimes a dozen or more of large Alligators in the evening, prepares his fire in the woods, where he has erected a camp for the purpose, and by morning has the oil rendered.

"I have frquently been very much amused when fishing in a bayou, where Alligators were numerous, by throwing a blown bladder on the water towards the nearest to me. The Alligator makes for it, flaps it towards its mouth, or attempts seizing it at once, but all in vain. The light bladder slides off; in a few minutes many Alligators are trying to seize this, and their evolutions are quite interesting. They then put one in mind of a crowd of boys running after a football. A black bottle is sometimes thrown also, tightly corked; but the Alligator seizes this easily, and you hear the glass give way under its teeth as if ground in a coarse mill. They are easily caught by negroes, who most expertly throw a rope over their heads when swimming close to shore, and haul them out instantly.

"But, my dear sirs, you must not conclude that Alligators are always thus easily conquered: there is a season when they are dreadfully dangerous; it is during spring, during the love season. The waters have again submerged the low countries; fish are difficult of access; the greater portion of the game has left for the northern latitudes; the quadrupeds have retired to the highlands; and the heat of passion, joined to the difficulty of procuring food, render these animals now ferocious, and very considerably more active. The males have dreadful fights together, both in the water and on the land. Their strength and weight adding much to their present courage, exhibit them like colossuses wrestling. At this time no man swims or wades among them; they are usually left alone at this season.

"About the first days of June the female prepares a nest; a place is chosen forty or fifty yards from the water, in thick bramble or cane, and she gathers leaves, sticks, and rubbish of all kinds, to form a bed to deposit her eggs; she carries the materials in her mouth, as a hog does straw. As soon as a proper nest is finished, she lays about ten eggs, then covers them with more rubbish and mud, and goes on depositing in different layers, until fifty or sixty or more eggs are laid. The whole is then covered up, matted and tangled with long grasses, in such a manner that it is very difficult to break it up. These eggs are the size of that of a goose, more elongated, and instead of being contained in a shell, are in a bladder, or thin transparent parchmentlike substance, yielding to the pressure of the fingers, yet resuming its shape at once, like the eggs of snakes and tortoises. They are not eaten even by hogs. The female now keeps watch near the spot, and is very wary and ferocious, going to the water from time to time only for food. Her nest is easily discovered, as she always goes and returns the same way, and forms quite a path by the dragging of her heavy body. The heat of the nest, from its forming a mass of putrescent manure, causes the hatching of the eggs; not that of the sun, as is usually believed.

"Some European writers say, that, at this juncture, the vultures feed on the eggs, and thereby put a stop to the increase of those animals. In the United States, I assure you, it is not so, nor can it be so, were the vultures ever so anxiously inclined; for, as I have told you before, the nest is so hard, and matted and plastered together, that a man needs his superior strength, with a strong sharp stick, to demolish it. The little Alligators, as soon as hatched, (and they all break shell within a few hours from the first to the last), force themselves through, and issue forth all beautiful, lively, and as brisk as lizards. The female leads them to the lake, but more frequently into small detached bayous for security sake, for now the males if they can get at them devour them by hundreds, and the wood ibis and the sandhill cranes also feast on them.

"I believe that the growth of Alligators takes place very slowly, and that an Alligator of twelve feet long, for instance, will most probably be fifty or more years old. My reasons for believing this to be fact is founded on many experiments, but I shall relate to you one made by my friend Bourgeat. That gentleman, anxious to send some Alligators as a present to an aquaintance in New York, had a bag of young ones quite small brought to his house.

They were put on the floor, to show the ladies how beautiful they were when young. One accidentally made its way out into a servant's room, and lodged itself snug from notice in an old shoe. The alligator was not missed, but, upwards of twelve months after this, it was discovered about the house, full of life, and apparently scarcely grown bigger; one of his brothers, that had been kept in a tub and fed plentifully, had grown only a few inches during the same period.

"Few animals emit a stronger odour than the Alligator; and, when it has arrived at a great size, you may easily discover one in the woods in passing fifty or sixty yards from it. This smell is highly musky, and so strong, that, when near, it becomes insufferable; but this I never experienced when the animal is in the water, although I have, whilst fishing, been so very close to them as to throw the cork of my fishing line on their heads, to tease them. In those that I have killed (and I assure you I have killed a great many), when opened, to see the contents of the stomach, or to take fresh fish out of them, I have regularly found round masses of a hard substance resembling petrified wood. These masses appeared useful to the animal in the process of digestion, like those found in the maws of some species of birds. I have broken some of them with a hammer, and found them brittle, and as hard as stones, which they outwardly resemble. And, as neither our lakes nor rivers, in the portion of the country I have hunted them in, afford even a pebble as large as a common egg, I have not been able to conceive how they are procured by the animals, if positively stones, or by what power wood can become stone in their stomachs."

Capt. Waterton, in his amusing book of Wanderings in South America, describes a most extraordinary adventure with an Alligator. It was first caught with a long iron hook, attached to a rope ; the Indians then drew him to the shore, and Waterton himself sprang upon his back,

seized his legs, and twisting them over his shoulders rode him safely up the bank, amid the shouts and exclamations of the savages.

THE SALAMANDER.

THE ancients have described a lizard that is bred from heat, that lives in the flames, and feeds upon fire, as its proper nourishment. It will be needless to say that there is no such animal existing ; and that, above all others, the modern Salamander has the smallest affinity to such an animal. The fact is, that, when the animal is exposed to fire, drops of a milky fluid ooze through all the pores of the skin. The same circumstance, however, occurs whenever it is handled. This fluid appears to be of an acrid nature.

There have been not less than seven sorts of this animal described by Seba; and to have some idea of the peculiarity of their figure, if we suppose the tail of a lizard applied to the body of a frog, we shall not be far from precision.

But it is not in figure that this animal chiefly differs from the rest of the lizard tribe. In conformation it is unlike, as the Salamander is produced alive from the body of its parent, and is completely formed the moment of its exclusion. It differs from them also in its general reputation of being venomous; no trials, however, that have been hitherto made, seem to confirm the truth of the report.

The Salamander best known in Europe, is from eight to eleven inches long; usually black, spotted with yellow; and, when taken in the hand, feeling cold to a great degree. There are several kinds. The black WATER NEWT is reckoned among the number. The idle report of its being inconsumable by fire, has caused many of these poor animals to be burnt; but we cannot say as philosophical martyrs; since scarce any philosopher would think it necessary to make the experiment. When thrown into the fire, the animal is seen to burst with the heat of its situation, and to eject its fluids. We are gravely told in the Philosophical Transactions, that this is a method the animal takes to extinguish the flames.

The whole of the lizard kind are so tenacious of life, that they will live several hours after the loss of the head: they also sustain the want of food in a surprising manner. One of them, brought from the Indies, lived nine months without any other food than what it received from licking a piece of earth, on which it was brought over;* another was kept by Seba, in an empty vial, for six months, without any nourishment; and Redi talks of a large one, brought from Africa, that lived for eight months, without taking any nourishment whatever. Indeed, as many of this kind, both Salamanders and lizards, are torpid, or nearly so, during the winter, the loss of their appetite for so long a time is the less surprising. If wetted with vin-

* Phil. Trans. ann. 1661. N. 21. Art. 7.

egar, however, or sprinkled with powdered salt, the animal soon dies in convulsions.

Directly descending from the crocodile, in this class, we find the CORDYLE, the TOCKAY, and the TEJUGUACU, all growing less in the order in which they are named. These fill up the chasm to be found between the crocodile and the African iguana.

THE IGUANA,

WHICH deserves our notice, is about three feet long, and the body about as thick as one's thigh. The skin is covered with small scales, like those of a serpent; and the back is furnished with a row of prickles that stand up, like the teeth of a saw. Both the jaws are full of very sharp teeth, and the bite is dangerous, though not venomous. The male has a skin hanging under his throat, which reaches down to his breast; and when displeased he puffs it up like a bladder: he is one-third larger and stronger than the female, though the strength of either avails them little towards their defence. The males are ash coloured, and the females are green.

The flesh of these may be considered as the greatest delicacy of Africa and America; and the sportsmen of those climates go out to hunt the Iguana, as we do in pursuit of the pheasant or the hare. In the beginning of the season, when the great floods of the tropical climates are passed away, and vegetation starts into universal verdure, the sportsmen are seen, with a noose and a stick, wandering along the sides of the rivers, to take the Iguana. This animal, though apparently formed for combat, is the most harmless creature of all the forest; it lives among the trees, or sports in the water, without ever offering to offend. There, having fed upon the flowers of the mahot, and the leaves of the mapou, that grow along the banks of the stream, it goes to repose upon the branches of the trees that hang over the water. Upon land, the animal is swift of foot; but when once in possession of a tree, it seems conscious of the security of its situation, and never offers to stir. There the sportsman easily finds it, and as easily fastens his noose round its neck. If the head be placed in such a manner that the noose cannot readily be fastened, by hitting the animal a blow on the nose with a stick, it lifts the head, and offers it in some measure to the noose. In this manner, and also by the tail, the Iguana is dragged from the trees, and killed by repeated blows on the head.

THE CHAMELEON

Is a very different animal; and as the iguana satisfies the appetites of the epicure, this is rather the feast of the philosopher. It was once, very unphilosophically, believed to have the power of subsisting on air alone. Like the crocodile, this little animal proceeds from an egg; and it also nearly resembles that formidable creature in form. It is found in all the warm countries, both of the old and the new world.

The head of a large Chameleon is almost two inches long; and thence to the beginning of the tail, four and a half; the tail is five inches long, and the feet two and a half; the thickness of the body is different at different times; for sometimes, from the back to the belly it is two inches; and sometimes but one; for it can blow itself up, and contract itself at pleasure. This swelling and contraction is not only of the back and belly, but of the legs and tail.

The Chameleon has the power of driving the air it breathes over every part of the body: however, it only gets between the skin and the muscles; for the muscles themselves are never swollen. The skin is very cold to the touch; and though the animal seems so lean, there is no feeling the beating of the heart. The surface of the skin is unequal, and has a grain not unlike shagreen, but very soft, because each eminence is as smooth as if it were polished. The colour of all these eminences, when the Chameleon is at rest in a shady place, is of a bluish gray; and the space between is of a pale red and yellow.

But when the animal is removed into the sun, then comes the wonderful part of its history. At first, it appears to suffer no change of colour, its grayish spots still continuing the same: but the whole surface soon seems to imbibe the rays of light; and the simple colouring of the body changes into a variety of beautiful hues. Wherever the light comes upon the body, it is of a tawny brown; but that part of the skin on which the sun does not shine changes into several brighter colours, pale yellow, or vivid crimson; which form spots of the size of half one's finger: some of these descend from the spine half way down the back; and others appear on the sides, arms, and tail. Sometimes the animal becomes all over spotted with brown spots, of a greenish cast. When it is wrapped up in a white linen cloth for two or three minutes, the natural colour becomes much lighter; but not quite white, as some authors have pretended; however, it must not hence be concluded that the Chameleon assumes the colour of the objects which it approaches; this is entirely an error, and probably has taken its rise from the continual changes it appears to undergo.

When the Chameleon changes place, and attempts to descend from an eminence, it moves with the utmost precaution, advancing one leg very deliberately before the other, still securing itself by holding whatever it can grasp by the tail. It seldom opens the mouth, except for fresh air; and, when that is supplied, discovers its satisfaction by its motions, and the frequent changes of its colour. The tongue is sometimes darted out after its prey, which are flies; and this is as long as the whole body. The eyes are remarkably little, though they stand out of the head; but the most extraordinary part of their conformation is, that the animal often moves one eye, when the other is entirely at rest; nay, sometimes one eye will seem to look directly forward, while the other looks backward; and one will look upwards, while the other regards the earth.

THE GREEN LIZARD.

THE colouring of this species is seen in its greatest brilliancy about the beginning of spring; when, after having thrown off its old covering, it exposes its new skin, with all its bright enamelled scales, to the genial warmth of the sun's rays, which, playing on the scales, gild them with undulating reflections. The upper parts of the body are of a beautiful green, more or less variegated with yellow, gray, brown, and even sometimes with red; the under parts being always more of a whitish colour. The colours of this species are subject to variety, becoming pale at certain seasons of the year, and more particularly after the death of the animal. It is chiefly in the warm countries that it shines with all its superb ornaments, like gold and precious stone. In those regions it grows to a larger size than in more temperate countries, being sometimes found thirty inches in length. The inhabitants of Africa eat the flesh of this animal. It is a gentle creature, and, if taken when young, may be rendered tame. If irritated, however, and driven to extremity, it will defend itself against a dog, and will fasten so firmly on his muzzle as to allow itself to be killed rather than forego its hold.

The Green Lizard is by no means confined to the warmest countries of both continents; it is found likewise in temperate regions, though it is there smaller and less numerous. It is not even unknown in Sweden, and in Kamtschatka; and in both countries, in spite of its beautiful appearance, it is looked on by the inhabitants with horror, from some strange superstitious prejudices. It is not an inhabitant of Europe.

To the class of Lizards we may refer the DRAGON, a most terrible animal, if we were to credit the invention of fable and superstition. Happily, however, such ravages are no where found to exist at present; and the whole race of Dragons is dwindled down to the flying Lizard, a little harmless creature, that only preys upon insects, and even seems to embellish the forest with its beauty.

THE SIREN

Is a creature not less extraordinary. It is the only biped

in this class of animals. In Carolina it is called the Mud Iguana, as it is chiefly found in muddy and swampy places.

THE TARANTULA

Is rather famous for the horror which it excites in the inhabitants of Italy, than for any other property. It is a deformed brown Lizard, somewhat thicker and rounder than other Lizards, and which, like the English *eft*, is found in old walls, or under the ruins of buildings. In perusing the accounts of naturalists and travellers, it will be necessary to observe the distinction between this animal and the spider which is called Tarantula, and of which we shall speak when we come to treat of insects.

The last animal of the Lizard kind we shall mention, is the CHALCIDIAN LIZARD, of Aldrovandus, very improperly called the Seps, by modern historians. This animal seems to mark the shade that separates the Lizard from the serpent race. It has four legs, like the Lizard; but so short, as to be utterly unserviceable in walking: it has a long slender body, like the serpent; and it is said to have the serpent's malignity also. These animals are found above three feet long, and thick in proportion, with a large head, and pointed snout. The whole body is covered with scales; and the belly is white, mixed with blue. It has four crooked teeth, as also a pointed tail, which, however, can inflict no wound. It is viviparous: upon the whole, it appears to bear a strong affinity to the viper; and, like that animal, its bite may be dangerous.

CHAP. XI.

In none of the countries of Europe is the Serpent tribe
sufficiently numerous to be truly terrible. The venomous

malignity also that has been ascribed to European Serpents
of old, is now utterly unknown; there are not above three
or four kinds that are dangerous, and the poison of all ope-
rates in the same manner. A burning pain in the part, easily removeable by timely applications, is the worst effect

that we experience from the bite of the most venomous Serpents of Europe.

Though, however, Europe be happily delivered from these reptiles, in the warm countries that lie within the tropics, as well as in the cold regions of the north, where the inhabitants are few, the Serpents propagate in equal proportion. All along the swampy banks of the rivers Niger and Oroonoko, where the sun is hot, the forests thick, and the men but few, the Serpents cling among the branches of the trees in infinite numbers, and carry on an unceasing war against all other animals in their vicinity. Travellers have assured us, that they have often seen large Snakes twining round the trunk of a tall tree, encompassing it like a wreath, and thus rising and descending at pleasure. In these countries, therefore, the Serpent is too formidable to become an object of curiosity, for it excites much more violent sensations.

We are not, therefore, to reject, as wholly fabulous, the accounts left us by the ancients of the terrible devastations committed by a single Serpent. It is probable, in early times, when the arts were little known, and mankind were but thinly scattered over the earth, that Serpents, continuing undisturbed possessors of the forest, grew to an amazing magnitude; and every other tribe of animals fell before them. We have many histories of antiquity, presenting us such a picture; and exhibiting a whole nation sinking under the ravages of a single Serpent. We are told, that while Regulus led his army along the banks of the river Bagrada, in Africa, an enormous Serpent disputed his passage over. We are assured by Pliny, who says that he himself saw the skin, that it was a hundred and twenty feet long, and that it had destroyed many of the army. At last, however, the battering engines were brought out against it; and these assailing it from a distance, it was soon destroyed.

With respect to their conformation, all Serpents have a very wide mouth, in proportion to the size of the head; and, what is very extraordinary, they can gape and swallow the head of another animal which is three times as big as their own. To explain this, it must be observed, that the jaws of this animal do not open as ours, in the manner of a pair of hinges, where bones are applied to bones, and play upon one another; on the contrary, the Serpent's jaws are held together at the roots by a stretching muscular skin; by which means they open as widely as the animal chooses to stretch them, and admit of a prey much thicker than the Snake's own body. The throat, like stretching leather, dilates to admit the morsel; the stomach receives it in part: and the rest remains in the gullet, till putrefaction and the juices of the Serpent's body unite to dissolve it.

As to the teeth, we shall speak more of them when we come to treat of the Viper's poison. The tongue in all these animals is long and forky. It is composed of two long fleshy substances, which terminate in sharp points, and are very pliable. Some of the Viper kind have tongues a fifth part of the length of their bodies; they are continually darting them out, but they are entirely harmless, and only terrify those who are ignorant of the real situation of their poison.

The skin is composed of a number of scales, united to each other by a transparent membrane, which grows harder as it grows older, until the animal changes it, which is generally done twice a year. This cover then bursts near the head, and the Serpent creeps from it, by an undulatory motion, in a new skin, much more vivid than the former. As the edges of the foremost scales lie over the ends of their following scales, so those edges, when the scales are erected, which the animal has a power of doing in a small degree, catch in the ground, like the nails in the wheel of a chariot, and so promote and facilitate the animal's progressive motion. The erecting these scales is by means of a multitude of distinct muscles, with which each is supplied, and one end of which is tacked to the middle of the foregoing.

This tribe of animals, like that of fishes, seems to have no bounds put to its growth: their bones are in a great measure cartilaginous, and they are consequently capable of great extension; the older, therefore, a Serpent becomes, the larger it grows; and as they seem to live to a great age, they arrive at an enormous size.

Leguat assures us, that he saw a Serpent in Java, that was fifty feet long; and Carli mentions their growing to above forty feet. Mr. Wentworth, who had large concerns at Berbice, in America, assures us, that in that country they grow to an enormous length. He one day sent out a soldier, with an Indian, to kill a wild fowl for the table; and they accordingly went some miles from the fort: in pursuing their game, the Indian, who generally marched before, beginning to tire, went to rest himself upon the fallen trunk of a tree, as he supposed it to be; but when he was just going to sit down, the enormous monster began to move, and the poor savage, perceiving that he had approached a Liboya, the greatest of all the Serpent kind, dropped down in an agony. The soldier, who perceived at some distance what had happened, levelled at the Serpent's head, and, by a lucky aim, shot it dead: however, he continued his fire, until he was assured that the animal was killed; and then, going up to rescue his companion, who was fallen motionless by its side, he, to his astonishment, found him dead likewise, being killed by the fright. Upon his return to the fort, and telling what had happened, Mr. Wentworth ordered the animal to be brought up, when it was measured, and found to be thirty-six feet long.

In the East Indies they grow also to an enormous size: particularly in the island of Java, where we are assured that one of them will destroy and devour a buffalo. In a letter printed in the German Ephemerides, we have an account of a combat between an enormous Serpent and a buffalo, by a person who assures us that he was himself a spectator. The Serpent had, for some time, been waiting near the brink of a pool, in expectation of its prey; when a buffalo was the first that offered. Having darted upon the affrighted animal, it instantly began to wrap it round with its voluminous twistings; and at every twist the bones of the buffalo were heard to crack almost as loud as the report of a cannon. It was in vain that the poor animal struggled and bellowed; its enormous enemy entwined it too closely to get free ; till at length, all its bones being mashed to pieces, like those of a malefactor on the wheel, and the whole body reduced to one uniform mass, the Serpent untwined its folds to swallow its prey at leisure. To prepare for this, and in order to make the body slip down the throat more readily, it was seen to lick the whole body over, and thus cover it with its mucus. It then began to swallow it at that end that offered least resistance ; while its length of body was dilated to receive its prey, and thus took in at once a morsel that was three times its own thickness. We are assured by travellers, that these animals are often found with the body of a stag in their gullet, while the horns, which they are unable to swallow, keep sticking out at their mouths.

But it is happy for mankind that the rapacity of these frightful creatures is often their punishment ; for whenever any of the Serpent kind have gorged themselves in this manner, whenever their body is seen particularly distended with food, they then become torpid, and may be approached and destroyed with safety.

Other creatures have a choice in their provision; but the Serpent indiscriminately preys upon all—the buffalo, the tiger, and the gazelle. One would think that the porcupine's quills might be sufficient to protect it; but whatever has life serves to appease the hunger of these devouring creatures: porcupines, with all their quills, have frequently been found in their stomachs when killed and opened; nay, they most frequently are seen to devour each other.

But though these animals are, above all others, the most voracious; and though the morsel which they swallow without chewing is greater than what any other creature, either by land or water, the whale itself not excepted, can devour, yet no animals upon earth bear abstinence so long as they. A single meal, with many of the snake kind, seems to be the adventure of a season; and is an occurrence for which they have been for weeks, nay, sometimes for months, in patient expectation. Their prey continues, for a long time, partly in the stomach, partly in the gullet; and a part is often seen hanging out of the mouth. In this manner it digests by degrees; and in proportion as the part below is dissolved, the part above is taken in. It is not therefore till this tedious operation is entirely performed, that the Serpent renews its appetite and its activity. But should any accident prevent it from issuing once more from its cell, it still can continue to bear famine for weeks, months, nay, for years together. Vipers are often kept in boxes for six or eight months, without any food whatever: and there are little Serpents sometimes sent over to Europe, from Grand Cairo, that live for several years in glasses, and never eat at all, nor even stain the glass with their excrement. Thus the Serpent tribe unite in themselves two very opposite qualities; wonderful abstinence, and yet incredible rapacity.

Though all Serpents are amphibious, some are much fonder of the water than others; and, though destitute of fins or gills, remain at the bottom, or swim along the surface with great ease. They can, however, endure to live in fresh water only; for salt is an effectual bane to the whole tribe.

Some Serpents have a most horrible fœtor attending them, which is alone capable of intimidating the brave. This proceeds from two glands near the vent, like those in the weasel or the polecat; and, like those animals, in proportion as they are excited by rage or by fear, the scent grows stronger. It would seem, however, that such Serpents as are most venomous are least offensive in this particular; since the Rattlesnake and the Viper have no smell whatever: nay, we are told that at Calcutta and Cranganore, in the East Indies, there are some very noxious Serpents, who are so far from being disagreeable, that their excrements are sought after, and kept as the most pleasing perfume. The Esculapian Serpent is also of this number.

Some Serpents bring forth their young alive; as the Viper: some bring forth eggs, which are hatched by the heat of their situation; as the common black Snake, and the majority of the Serpent tribe. When a reader, ignorant of anatomy, is told that some of these animals produce their young alive, and that some produce eggs only, he is apt to suppose a very great difference in their internal conformation, which makes such a variety in their manner of bringing forth. But this is not the case: these animals are internally alike, in whatever manner they produce their young; and the variety in their bringing forth is rather a slight than a real discrimination. The only difference is, that the Viper hatches her eggs, and brings them to maturity within her body; the Snake is more premature in her productions, and sends her eggs into the light some time before the young ones are capable of leaving the shell. Thus, if either are opened, the eggs will be found in the womb, covered with their membranous shell, and adhering to each other, like large beads on a string. In the eggs of both, the young ones will be found, though at different stages of maturity: those of the Viper will crawl and bite the moment the shell that encloses them is broken open; those of the Snake are not yet arrived at their perfect form.

Father Labat took a Serpent of the Viper kind, that was nine feet long, and ordered it to be opened in his presence. He then saw the manner in which the eggs of these animals lie in the womb. In this creature there were six eggs, each of the size of a goose egg, but longer, more pointed, and covered with a membranous skin, by which also they were united to each other. Each of these eggs contained from thirteen to fifteen young ones, about six inches long, and as thick as a goose quill. These little mischievous animals were no sooner let loose from the shell, than they crept about, and put themselves into a threatening posture, coiling themselves up, and biting the stick with which he was destroying them. In this manner he killed seventy-four young ones; those that were contained in one of the eggs escaped at the place where the female was killed, by the bursting of the egg, and their getting among the bushes.

The last distinction that we shall mention, but the most material among Serpents, is, that some are venomous and some inoffensive; but not above a tenth of their number are actually venomous.

From these noxious qualities in the Serpent kind, it is no wonder that not only man, but beasts and birds, carry on an unceasing war against them. The ichneumon of the Indians, and the peccary of America, destroy them in great numbers. These animals have the art of seizing them near the head; and it is said that they can skin them with great dexterity. The vulture and the eagle also prey upon them in great abundance; and often, sousing down from the clouds, drop upon a long Serpent, which they snatch struggling and writhing in the air. Dogs also are bred up to oppose them.

Father Feuillée tells us that, being in the woods of Martinico, he was attacked by a large Serpent, which he could not easily avoid, when his dog immediately came to his relief, and seized the assailant with great courage. The Serpent entwined him, and pressed him so violently, that the blood came out of his mouth, and yet the dog never ceased till he had torn it in pieces. The dog was not sensible of his wounds during the fight; but, soon after, his head swelled prodigiously, and he lay on the ground as dead. But his master having found, hard by, a banana tree, he applied its juice, mixed with treacle, to the wound; which recovered the dog, and quickly healed his sores.

In India there is nothing so common as dancing Serpents, which are carried about in a broad flat vessel somewhat resembling a sieve. These erect and put themselves in motion at the word of command. When their keeper sings a slow tune, they seem by their heads to keep time; when he sings a quicker measure, they appear to move more brisk and lively. All animals have a certain degree of docility; and we find that Serpents themselves can be brought to move and approach at the voice of their master. From this trick successfully practised before the ignorant, it is most probable has arisen all the boasted pretensions which some have made to charming of Serpents; an art to which the native Americans pretend at this very day.

VENOMOUS SERPENTS.

In all countries, the poison of the Serpent is sufficiently
formidable to deserve notice, and to excite our attention
to its nature and effects. It will therefore, in the first
place, be proper to describe its seat in the animal, as also
the instrument by which the wound is made and the poison
injected.   In all this venomous class of reptiles, whether
the Viper, the Rattlesnake, or the Cobra di Capello, there
are two large teeth or fangs that issue from the upper jaw,
and that hang out beyond the lower. The rest of the
Snake tribe are destitute of these; and it is most probable
that wherever these fangs are wanting the animal is harmless; on the contrary, wherever they are found, it is to be
avoided as the most pestilent enemy.   Our first great attention, therefore, upon seeing a Serpent, should be directed to the teeth.   The Black Snake, the Liboya, the
Blind Worm, and a hundred others that might be mentioned, have their teeth of an equal size, fixed into the jaws,
and with no other apparatus for inflicting a dangerous
wound than a dog or a lizard; but it is otherwise with the
venomous tribe we are now describing: these are well
furnished, not only with a laboratory, where the poison is
formed, but a canal, by which it is conducted to the jaw;
a bag under the tooth for keeping it ready for every occasion; and also an aperture in the tooth itself for inject-
ing it into the wound. The venom contained in this bag
is a yellowish, thick, tasteless liquor, which injected into
the blood is death, yet which may be swallowed without
any danger.

The fangs that give the wound are large in proportion
to the size of the animal that bears them; crooked, yet
sharp enough to inflict a ready wound. They grow one
on each side, and sometimes two, from two moveable

N3

bones in the upper jaw, which, by sliding backward or forward, have a power of erecting or depressing the teeth at pleasure. In these bones are also fixed many teeth, but no way venomous, and only serving to take and hold the animal's prey. If a Viper inflicts the wound, and the remedy be neglected, the symptoms are not without danger. It first causes an acute pain in the place affected, attended with a swelling, first red, and afterwards livid. To this succeed great sickness in the stomach, bilious and convulsive vomitings, cold sweats, pains about the navel, and death itself. These symptoms are much more violent, and succeed each other more rapidly, after the bite of a Rattlesnake; but when the person is bit by the Cobra di Capello, he dies in an hour, his whole frame being dissolved into a putrid mass of corruption.

In the Eastern and Western Indies, the number of noxious Serpents is various; in England, the inhabitants are acquainted only with one. The Viper is the only animal of Great Britain whose bite is dangerous. In the tropical climates, the Rattlesnake, the Whip Snake, and the Cobra di Capello, are the most formidable, though by no means the most common.

THE COMMON EUROPEAN VIPER.

THE Common Viper is a native of many parts of Europe; but the dry, stony, and in particular the chalky countries abound with them. In the East Indies also it is found, and varies very slightly from that which belongs to Europe. It can equally support the vicissitudes of very cold climates, it being an inhabitant of Sweden, where its bite is nearly as dangerous as in the warmer regions of Europe; and likewise of Russia, and of several parts of Siberia, where it is very numerous, as the superstitions of the people deter them from endeavoring to destroy this noxious reptile, because that they conceive some terrible disaster will follow the attempt. This animal seldom grows to a greater length than two feet; though sometimes they are found above three. The ground colour of their bodies is a dirty yellow; that of the female is deeper.

The back is marked the whole length with a series of rhomboid black spots, touching each other at the points; the sides with triangular ones; the belly entirely black. It is chiefly distinguished from the common ringed Snake by the colour, which in the latter is more beautifully mottled; as well as by the head, which is thicker than the body; but particularly by the tail, which, in the Viper, though it ends in a point, does not run tapering to so great a length as in the other. When, therefore, other distinctions fail, the difference of the tail can be discerned at a single glance.

The Viper differs from most other Serpents in being much slower, as also in excluding its young completely formed, and bringing them forth alive. The kindness of Providence seems exerted not only in diminishing the speed, but also the fertility, of this dangerous creature. They couple in May, and are supposed to be about three months before they bring forth, and have seldom above eleven eggs at a time. These are of the size of a blackbird's egg, and chained together in the womb like a string of beads. Each egg contains from one to four young ones; so that the whole of a brood may amount to about twenty or thirty. They continue in the womb till they come to such perfection as to be able to burst from the shell; and they are said by their own efforts to creep from their confinement into the open air, where they continue for several days without taking any food whatsoever. When they are in danger, they retreat into the mouth o the mother.

The Viper is capable of supporting very long abstinence, it being known that some have been kept in a box six months without food; yet during the whole time they did not abate of their vivacity. They feed only a small part of the year, but never during their confinement; for if mice, their favourite diet, should at that time be thrown into their box, though they will kill, yet they will never eat them. When at liberty, they remain torpid throughout the winter; yet, when confined, have never been observed to take their annual repose.

Vipers crawl slowly at all times, and in general only attack such smaller animals as are their usual food. They never willingly assail man or the larger animals, except when wounded, trampled on, or irritated, when they become furious, and are apt to bite severely. From the firmer manner in which their spine is articulated, they are unable to twist themselves round so much as most other Serpents. It is only the head that they can turn with any considerable degree of agility; owing to this circumstance, they are easily taken. For this purpose, some persons use a forked stick, to fix the Viper by the neck; and then, seizing it by the tail, throw it into a bag. Others, holding down its head with the end of a stick, take it fast in the left hand by the neck; and while the animal makes ineffectual efforts to defend itself, with its mouth wide open, cut out its poisonous fangs with a pair of scissors or a knife: the Viper, after this, is incapable of doing injury, and may be handled with perfect safety. The English Viper-catchers have the boldness to seize them suddenly by the neck, or even by the tail, with their hands; and holding them with a firm grasp, the animal is altogether incapable of turning itself sufficiently to bite the hand that holds it fast.

THE EGYPTIAN VIPER.

THIS is said to be the Officinal Viper of the Egyptians, and is by some supposed to be the Asp of Cleopatra, by the bite of which that high-spirited princess determined to die, rather than submit to be carried to Rome in order to grace the triumph of Augustus. It is imported in considerable quantities every year to Venice, for the use of the apothecaries in the composition of theriaca, and for other purposes. It is abundant in Egypt; and is found in other parts of Africa as well as in Asia. It is from twenty inches long to three feet and upwards, variegated with rich chestnut brown spots or bands, on a lighter brown ground, the scales remarkably short, close set, and hard; the eyes are vertical; the head compressed, and covered with very minute dark brown scales, and reddish stripes. It is very

N5

poisonous, but not often fatally so. The death produced by its bite is said to be speedy, but devoid of pain.

THE EGYPTIAN ASP.

THIS animal is about three feet in length; the head is rather large, and covered with small carinated scales, the body with larger, of similar structure: the colour is pale rufous gray, and along the upper parts are three longitudinal ranges of deep rufous spots, bordered with black, and which unite or become confluent towards the tail, in such a manner as to exhibit the appearance of a zigzag band, resembling in some degree that of the common Viper: the under parts of a dusky colour, marbled with dull yellow: in the structure of its fangs it resembles the Viper, and is said to be equally poisonous. The nose is terminated by an erect wart-like excrescence. It preys by smell, and feeds on rats, mice, lizards, frogs, toads, &c.

THE WATER VIPER.

THIS animal is a native of the Carolinas, and the vicinity, and, next to the rattle snake, is the largest species of Serpent in that country. It is of an ash colour, variegated with yellow spots. Unlike other Vipers, it is very active, and it catches fish with great dexterity. In summer, numbers of them are seen lying on the branches of trees which hang over rivers, waiting to surprise birds or fish, and they often drop into boats on the heads of the men. At the end of its tail is a horny and harmless substance, concerning the fatal powers of which many absurd stories have been told. Its fang bite is said, however, to be as venomous as that of the rattlesnake.

THE HORNED VIPER.

THE Cerastes, or Horned Viper, which commonly grows to the length of about a foot or fifteen inches, but sometimes more than two feet, is distinguished by a pair of horns or curved processes, situated above the eyes, and pointing forward: these horns have nothing analogous in their structure to the horns of quadrupeds, and are by no means to be considered in the light of either offensive or defensive weapons; they are moveable, and about one-sixth of an inch long.

The head of the Cerastes is flattened; the muzzle thick and short; the eye yellowish green. The hind part of the head is narrower than the part of the body to which it is attached. The scales of the head are of the same size with those on the back, or only a little less; and all the scales are oval with a longitudinal ridge. The general colour of the back is yellowish, with irregular spots of different degrees of darkness, in form of transverse bands. The under surface of the body is brighter. The Cerastes inhabits the greatest part of the eastern continent, especially the desert sandy part of it. It abounds in the three Arabias, and in Africa. In Egypt it appears to be partially domesticated, as it will enter the houses when the family

are at table, pick up the scattered crumbs, and retire without doing injury to any one.

This animal can endure hunger and thirst much longer
than most Serpents; some naturalists assert that it can
exist five years without nourishment. But, though able to
live long without food, these animals are extremely voracious, and attack small birds, quadrupeds, and reptiles, with
great eagerness. Their skin is capable of very great extension, and they can swallow food till increased to twice their
ordinary size; and, as their digestion is extremely slow,
they become, when gorged, quite torpid and motionless,
in which situation they are very easily killed.

THE LIFE-CONSUMING VIPER.

This venomous Serpent well deserves the above name;
and is most deservedly dedicated to one of the fates, on
account of the violent poison which distils from its mur-
derous fangs. It is a native of the burning zone of Africa,
and of the warm regions of South America: it is also found
in the island of Ceylon, and in the remote provinces of
Asia. The ground colour is a silvery white, variegated
with black irregular spots and blotches. The head is
broad, and the mouth large and blunt; the eyes are like
drops of pearl, surrounded with a green iris; the head is
cordated with quadruple rows of transverse white bead-like
lines, passing behind the eyes, on a black ground; the
scales are varied, some large and shield-formed, others
small and pointed, and all remarkably strong and close set.
This is a most dangerous and formidable Serpent; its poison, though not so rapid in its effects as some, yet causes
a stupefaction of the mental faculties, a wasting of the
flesh by incurable consumption, and finally death.

THE FATAL VIPER.

THIS animal has a short round head, and a large wide mouth, armed with four curvated teeth, two in each jaw; its eyes dart fire; and its bite inflicts cruel and inevitable death. It has also the faculty of erecting its scales at pleasure, or when irritated, and of closing them again with a rattling noise. A border of silver white scales surrounds the mouth; the tongue is fleshy and forked, which the creature can protrude to a great length when offended; at which time it also shows its teeth in a menacing posture, like a snarling dog, and thus it can show or conceal its fangs at pleasure. The scales on the upper part of the body are elegantly speckled with pale yellow, cinereous gray, black, brown, and white, glistening most superbly in the sun. It is a native of South America, and of the island of Ceylon; it will attack man or beast with great fury, erecting its crest, and darting forward with singular rapidity and courage. The male is deeper coloured than the female, and appears to have a larger body and a more slender tail. The general length of this Snake seems to be about four or five feet, and the tail is long in proportion to the body.

THE ANNULATED SNAKE.

THE ground of this Snake is white, with brown transverse bars, which are straight and distinct on the back, but run into one another on the belly. The tail is slender, and has two ranges of imbricated scales on its under surface. When irritated, or preparing to bite, it raises up the fore part of its body, and carries its head in a bending position. It is a native of South America, and there are a great number of elegant varieties of it, particularly the large Annulated Snake of Surinam.

THE RATTLESNAKE

Is bred in North and South America, and in no part of the Old World. Some are as thick as a man's arm, and six feet in length; but the most usual size is from four to five feet long. In most particulars it resembles the viper: it differs, however, in having a large scale, which hangs like a penthouse over each eye. They are of an orange tawny, and blackish colour on the back; and of an ash colour on the belly, inclining to lead. The male may be readily distinguished from the female, by a black velvet spot on the head, and by the head being smaller and longer. But that which, besides their superior malignity, distinguishes them from all other animals, is their rattle, an instrument lodged in their tail, by which they make such a loud, rattling noise, when they move, that their approach may readily be perceived, and the danger avoided. This rattle, which is placed in the tail, somewhat resembles, when taken from the body, the curb chain of a bridle: it is composed of several thin, hard, hollow bones, linked on each other, and rattling upon the slightest motion. It is supposed by some that the Snake acquires an additional bone every year; and that from this its age may be precisely known: however this may be, certain it is that the young Snakes of a year or two old have no rattles at all; while many old ones have been killed, that had from eleven to thirteen joints each. They shake and make a noise with these rattles with prodigious quickness when they are disturbed; however, the peccary and the vulture are no way terrified at the sound, but hasten at the signal to seize the Snake, as their most favourite prey.

It is very different with almost every other animal. The certain death which ensues from this terrible creature's bite makes a solitude wherever it is heard. It moves along most majestically; neither seeking to offend the larger animals, nor fearing their insults. If unprovoked, it never meddles with any thing but its natural prey; but when accidentally trodden upon, or pursued to be destroyed, it then makes a dreadful and desperate defence. It erects itself upon its tail, throws back its head, and inflicts the wound in a moment; then parts, and inflicts a second wound: after which, we are told by some, that it remains torpid and inactive, without even attempting to escape.

The very instant the wound is inflicted, though small in itself, it appears more painful than the sting of a bee. This pain, which is so suddenly felt, far from abating, grows every moment more excruciating and dangerous: the limb swells; the venom reaches the head, which is soon of a monstrous size; the eyes are red and fiery; the heart beats quick, with frequent interruptions: the pain becomes insupportable, and some expire under it in five or six hours;

but others, who are of stronger constitutions, survive the agony for a few hours longer, only to sink under a general mortification which ensues, and corrupts the whole body.

The usual motion of the Rattlesnake is with its head to the ground. When, however, it is alarmed, it coils its body into a circle, with its head erect, and its eyes flaming in a terrific manner. But it cannot pursue rapidly, and has no power of springing on its enemy.

Rattlesnakes are viviparous, producing their young, generally about twelve in number, in the month of June, and by September these acquire the length of twelve inches. It has been well attested that they adopt the same mode of preserving their young from danger as that attributed to the common viper, receiving them into their mouth and swallowing them. It is believed by some naturalists to have the power of fascinating its prey by gazing at it, so as to render it incapable of flight; but others are doubtful as to this being a fact. The probability seems to be, that the victim is prevented from escaping merely by the extreme terror which its formidable enemy inspires.

A Serpent, called the WHIP SNAKE, is still more venomous than the former. This animal, which is a native of the East, is about five feet long, yet not much thicker than the thong of a coachman's whip. It is exceedingly venomous; and its bite is said to kill in about six hours. One of the Jesuit Missionaries, happening to enter an Indian pagoda, saw what he took to be a whipcord lying on the floor, and stooped to take it up; but upon handling it, what was his surprise to find that it was animated, and no other than the Whip Snake, of which he had heard such formidable accounts. Fortune, however, seemed favourable to him; for he grasped it by the head, so that it had no power to bite him, and only twisted its folds up his arm.

In this manner he held it, till it was killed by those who came to his assistance.

To this formidable class might be added, the JACULUS of Jamaica, one of the swiftest of the serpent kind. The HÆMORRHOIS, so called from the hæmorrhages which its bite is said to produce; the SEPS, whose wound is very venomous, and causes the part affected to corrupt in a very short time; the CORAL SERPENT, which is red, and whose bite is said to be fatal. But the COBRA DI CAPELLO, or HOODED SERPENT, inflicts the most deadly and incurable wounds. Of this formidable creature there are five or six different kinds; but they are all equally dangerous, and their bite is followed by speedy and certain death. It is from three to eight feet long, with two large fangs hanging out of the upper jaw. It has a broad neck, and a mark of dark brown on the forehead; which, when viewed front-wise, looks like a pair of spectacles; but behind, like the head of a cat. The eyes are fierce, and full of fire; the head is small, and the nose flat, though covered with very large scales, of a yellowish ash colour; the skin is white; and the large tumour on the neck is flat, and covered with oblong, smooth scales.

SERPENTS WITHOUT VENOM.

THIS class of Serpents all want that natural mechanism by which the poisonous tribe inflict such deadly wounds: they have no glands in the head for preparing venom; no conduits for conveying it to the teeth; no receptacles there; no hollow in the instrument that inflicts the wound. Their bite, when the teeth happen to be large enough to penetrate the skin, for in general they are too small for this purpose, is attended with no other symptoms than those of an ordinary puncture; and many of this tribe, as if sensible of their own impotence, cannot be provoked to bite, though ever so rudely assaulted. They hiss, dart out their forky tongues, erect themselves on the tail, and call up all their terrors to intimidate their aggressors, but seem to consider their teeth as unnecessary instruments of defence, and never attempt to use them. Even among the largest of this kind, the teeth are never employed in the most desperate engagements. When a hare or a bird is caught, the teeth may serve to prevent such small game from escaping; but, when a buffalo or a tiger is to be encountered, it is by the strong folds of the body, by the fierce verberations of the tail, that the enemy is destroyed: by thus twining round, and drawing the knot with convulsive energy, this enormous reptile breaks every bone in the quadruped's body, and then at one morsel devours its prey.

Hence we may distinguish the unvenomous tribe into two kinds: first, into those which are seldom found of any considerable magnitude, and that never offend animals larger and more powerful than themselves, but which find their chief protection in flight, or in the doubtfulness of their form; secondly, into such as grow to an enormous size, fear no enemy, but indiscriminately attack all other animals, and devour them. Of the first kind is the Common Ringed Snake, the Blind Worm, the Esculapian Serpent, the Amphisbæna, and several others. Of the second, the Boas, the Anaconda, and the Depona.

THE BLACK OR RINGED SNAKE

Is the largest of English Serpents, sometimes exceeding four feet in length. The neck is slender, the middle of the body thick, the back and sides covered with small scales; the belly with oblong, narrow, transverse plates; the colour of the back and sides is of a dusky brown; the middle of the back marked with two rows of small black spots, running from the head to the tail; the plates on the belly are dusky; the scales on the sides are of a bluish white; the teeth are small and serrated, lying on each side

of the jaw, in two rows. The whole species is perfectly inoffensive, taking shelter in dunghills, and among bushes in moist places: whence they seldom remove, unless in the midst of the day, in summer, when they are invited out by the heat to bask themselves in the sun.

This Snake preys upon frogs, insects, worms, mice, and young birds; and, considering the smallness of the neck, it is amazing how large an animal it will swallow. It is said to be particularly fond of milk.

The BLACK SNAKE of the United States, which is larger than the above, and generally grows to six feet long; takes a prey proportionable to its size—squirrels and small birds. It is sometimes found in the neighbourhood of the hen-roost, and will devour the eggs, even while the hen is sitting upon them. But its usual haunts are meadows and dry thickets. It may be often seen among whortle-berry bushes, waiting to make a prey of the birds that are hopping among them. Its colour is a glossy black, some-times tinged with blue. It seizes its prey with great quick-ness, and kills it by coiling round the body, in the man-ner of the boa constrictor. It climbs trees with facility, and in general is more disposed to seek safety by conceal-

ment than flight. When it adopts the latter course, its
speed is very great. Instances have been related, which
seem to prove that it is capable of outstripping a dog.
It is perfectly harmless, and generally seeks an instant
retreat when approached by man. We have heard stories
of its coiling about people; but there is no well authenticated account within our knowledge, of its ever having
done harm to any one.

THE STRIPED SNAKE

Is familiarly known to every body in this country. It is
often seen by the road side, about gardens and stone walls.
It lives upon frogs and toads, which it swallows whole. It
seems to be one of the most innocent and harmless creatures in existence, and nothing but the common prejudice
against serpents can account for the war that is waged
against this timid animal. The doom of Cain is upon it,
for whoever meets it, slays it.

The whole of this tribe of serpents are oviparous, laying
eighty or a hundred eggs at a time, in dunghills or hot-
beds: the heat of which, aided by that of the sun, brings
them to maturity. During winter, they lie torpid, in banks,
or hedges, and under old trees.

THE BLIND WORM

Is another harmless reptile, with a formidable appearance.
The usual length of this species is eleven inches. The
eyes are red, the head small, the neck still more slender;
from that part the body grows suddenly, and continues of
an equal bulk to the tail, which ends quite blunt. The
colour of the back is cinereous, marked with very small
lines composed of minute black specks. The motion of
this Serpent is slow; from which, and from the smallness
of the eyes, are derived its name; some calling it the Slow,

and some the Blind Worm. Like all the rest of the kind, in our climates, they lie torpid during the winter, and are sometimes found, in vast numbers, twisted together. This animal, like the former, is perfectly innocent; like the viper, however, it brings forth its young alive.

THE SOOTY AMPHISBÆNA, OR BLIND SNAKE.

THIS species grows usually to the length of one or two feet, of which the tail never exceeds an inch, or an inch and a quarter. It is from the extremities of its body being of an equal thickness that it derives its name. The eyes are exceedingly small, and covered in such a manner by a membrane, as to be hardly perceptible; from which circumstance, it has been called the Blind Serpent. The top of the head is covered by six large scales, in three rows of two each; and the body is entirely covered with smooth scales of an almost square form, arranged in regular transverse rings. It has a hard skin of an earthy colour.

This animal is found in India, particularly the isle of Ceylon; and likewise in South America. Its habits are in a great measure unknown; but it feeds on earthworms, beetles, and various insects: it is particularly fond of devouring ants, which in numberless legions often destroy all before them, leaving every thing desolate as if destroyed by fire. Having the power of advancing or withdrawing itself without injury, in consequence of its structure, this Serpent is peculiarly fitted for penetrating into the subterraneous retreats of ants, worms, and other insects; and is able to dig deeper than any other Serpent, its skin being very hard, and its muscles very strong. The Amphisbæna is not venomous.

THE AQUATIC VIPER.

THIS species is a native of India, where it frequents wet swampy fields, and is commonly reckoned a water Snake; it is about two feet nine inches in length, and in circumference three inches and a half: the head is rather broad, somewhat depressed, and laterally compressed: its body is covered with large scales, gradually diminishing towards the tail, which is eleven inches in length, slightly carinated, tapering very gradually, and terminating sharply. The head is of a dusky hue; the rest of the animal a yellowish brown, with numerous round black spots, joined by narrow fillets regularly disposed in oblique rows, a few scales of light yellow being interpersed: the abdomen yellowish white.

A Viper of this kind was caught in the Lake of Ankapilly, in one of the traps employed for catching eels; and though several experiments were made with a stick to try to irritate it, it did not either hiss or snap; neither was it provoked to bite a chicken, though pecked several times by the animal: while it lay coiled up, a chicken properly secured was laid upon it, but it continued quiet, without attempting to wreath round the chicken, or otherwise to annoy it; and when the bird fluttered and struggled to get loose, the Snake, as if afraid, crept away. Its forbearance, however, might arise from its hunger being satisfied, as it had been feeding not long before the experiment. It is, however, certainly not venomous, does not appear to be irascible, and is considered by the natives to be harmless.

THE ESCULAPIAN SERPENT

Of Italy is among this number. It is there suffered to crawl about the chambers, and often gets into the beds where people lie. It is a yellow Serpent, of about an ell long; and, though innocent, yet will bite when exasperated. They are said to be great destroyers of mice; and this may be the reason why they are taken under human protection. The BOYUNA of Ceylon, is equally a favourite among the natives; and they consider the meeting it as a sign of good luck. The SURINAM SERPENT, which some improperly call the Ammodytes, is equally harmless and desirable among the savages of that part of the world. They consider themselves as extremely happy, if this animal comes into their huts. The colours of this Serpent are so many and beautiful, that they surpass all description; and these, perhaps, are the chief inducements to the savage, to consider its visits so very fortunate. A still greater favourite is the PRINCE of SERPENTS, a native of Japan, that has not its equal for beauty. The scales which cover the back are reddish, finely shaded, and marbled with large spots of irregular figures mixed with black. The fore part of the head is covered with yellow; the forehead marked with a black marbled streak, and the eyes handsome and lively. But the GERENDA of the East Indies is the most honoured and esteemed. To this animal, which is finely spotted with various colours, the natives of Calicut pay divine honours; and, while their deity lies coiled up, which is its usual posture, the people fall upon their faces before it, with stupid adoration. The African Gerenda is larger, and worshipped in the same manner, by the inhabitants of the coasts of Mozambique.

But in the larger tribe of Serpents there is nothing but danger to be apprehended. This formidable class, though without venom, have something frightful in their colour, as well as their size and form. They want that vivid hue, with which the savages are so much pleased in the lesser kinds. They are all found of a dusky colour, with large teeth, which are more formidable than dangerous.

THE GREAT BOA.

THE Snakes of the Boa tribe are distinguished from those of other tribes by the under surface of the tail being covered with scuta or undivided plates, like those on the belly, and by having no rattle. The individuals are strong but not venomous, never attack except openly, and from necessity, and conquer by dint of strength. Three species are found in Asia, the others belong to the warmer parts of the new continent.

The ground colour of the body of the Great Boa, which is the largest and strongest of the Serpent race, is yellowish gray, on which is distributed, along the back, a series of large, chain-like, reddish brown, and sometimes perfectly red variations, with other small and more irregular marks and spots.

The Great Boa is frequently from twentyfive to thirty feet in length, and of a proportionate thickness. The rapacity of these creatures is often their own destruction: for whenever they seize and swallow their prey, they seem like surfeited gluttons, unwieldy, stupid, helpless, and sleepy. They at the same time seek for some retreat, where they may lurk for several days together, and digest their meal in safety. The smallest effort will then destroy them: they scarcely can make any resistance: and, equally unqualified for flight or opposition, even the naked Indians do not fear to assail them. But it is otherwise when this sleeping interval of digestion is over; they then issue, with famished appetites, from their retreats, and with accumulated terrors, while every animal of the forest flies from their presence. One of them has been known to kill and devour a buffalo. Having darted upon the affrighted beast (says the narrator), the Serpent instantly began to wrap him round with its voluminous twistings; and at every twist the bones of the buffalo were heard to crack as loud as the report of a gun. It was in vain that the animal struggled and bellowed; its enormous enemy entwined it so closely, that at length all its bones were crushed to pieces, like those of a malefactor on the wheel, and the whole body was reduced to one uniform mass: the Serpent then untwined its folds, in order to swallow its prey at leisure. To prepare for this, and also to make it slip down the throat more smoothly, it was seen to lick the whole body over, and thus to cover it with a mucilaginous substance. It then began to swallow it, at the end that offered the least resistance; and in the act of swallowing, the throat suffered so great a dilatation, that it took in at once a substance that was thrice its own thickness. In

1799, a Malay seaman was almost instantaneously crushed to death, in the island of Celebes, by one of these Serpents, thirty feet in length, which seized him by the right wrist, and twined round his head, neck, breast and thigh.

THE INDIAN BOA.

THIS specimen of the Boa, which appears to be the Pedda Poda of Dr. Russell's Indian Serpents, is now to be seen in the Tower. It grows to the length of fifteen or sixteen feet. There are two hundred and fifty-two transverse plates on the under surface of the body, and sixty-two pairs of scales beneath the tail. The back is marked with a series of large circular brown blotches bordered with black; and along the sides are scatered numerous smaller spots. A yellowish brown, lighter beneath, is the ground colour.

There is a female in the Tower menagerie of London, which, not long since, produced a cluster of fourteen or fifteen eggs. None of them were hatched. The mother, however, "evinced the greatest anxiety for their preservation, coiling herself around them in the form of a cone, of which her head was the summit, and guarding them from external injury with truly maternal solicitude. They were visible only when she was occasionally roused; in which case she raised her head, which formed as it were the cover of the receptacle in which they were enclosed, but replaced it again as quickly as possible, allowing to the spectator only a momentary glance at her cherished treasure."

THE ANACONDA.

THIS name, which appears to be of Ceylonese origin, has been popularly applied to all the larger and more powerful Snakes. The animal (figured p. 316), which is now in the Tower, differs in no very remarkable points from the Indian Boa; the only distinctions between them consisting in the lighter colour, the greater comparative size of the head, and the acuteness of the tail, in the Anaconda.

Two snakes, a Boa and an Anaconda, about thirteen feet in length, have recently been exhibited in Boston and New York. On one occasion we saw two rabbits put into the cage, which jumped about for some time over the coiled serpents, without any apparent apprehension. They were slowly pursued by the heads of the reptiles, till at length, with a sudden dart, the rabbits were seized at the nose, and instantly their bodies were closely compressed by the coils which were wound around them. As the suffocated animals struggled in the agonies of death, the coils were drawn closer, until we could distinctly hear the bones of the rabbits broken and crushed to pieces. All who witnessed the exhibition, could easily believe the accounts which are given by travellers, of the prodigious force displayed by some of the great serpents of India and Africa.

THE DEPONA.

To this class of large Serpents we may refer the Depona, a native of Mexico, with a very large head, and great jaws. The mouth is armed with cutting, crooked teeth, among which there are two longer than the rest, placed in the fore part of the upper jaw, but very different from the fangs of the viper. All around the mouth there is a broad, scaly border; and the eyes are so large, that they give it a very terrible aspect. The forehead is covered with very large scales, on which are placed others, that are smaller, curiously ranged; those on the back are gray-ish. Each side of the belly is marbled with large square spots, of a chestnut colour; in the middle of which is a spot, which is round and yellow. They avoid the sight of man; and, consequently, never do much harm.

APPENDIX.

THE SEA-SERPENT.

It may be neither useless nor uninteresting to add here some account of the various evidences respecting the existence of a marine animal, called the Sea-Serpent. The following pages have been gathered partly from the newspapers, and partly from an unpublished pamphlet on the subject.

Our readers are probably aware of the description given of the Sea-Snake, which had been formerly 'seen on the coast of Norway, by Pontoppidan, Bishop of Bergen. It is not, perhaps, so well understood, that in a very old edition of the natural history of Norway, by that author, is an engraving of the animal, strictly conforming to the numerous drawings of him, as he has appeared upon this coast, with one exception. In the old book he is pictured with a kind of mane, hanging in the water, which has never been observed to belong to him in our seas—and probably the appearance was only occasioned by his rapid motion through the water, causing great foam round his head like that made by the bow of a vessel, when sailing very fast. Pontoppidan, in his account, says that these creatures make their appearance in the months of July and August, which is their spawning time, when they come to the surface in calm weather. If the wind raises the waves, they descend. Numerous persons, he observes, agree very well in the general description of the animal.—He most frequently appeared in the North Sea, near the Manor of Nordland—where all the inhabitants were not only convinced of his existence, but thought it surprising that the fact should be doubted ; and he confesses he questioned it himself, until his suspicion was removed by credible evidence. He further observes :—

"Though one cannot have an opportunity of taking the exact dimensions of this creature, yet all that have seen it are unanimous in affirming, as far as they can judge at, a distance, that it appears to be the length of a cable, i. e. one hundred fathoms, or six hundred English feet; that it lies on the surface of the water, when it is very calm, in many folds ; and that there are in a line with the head some small parts of the back to be seen above the surface of the water when it moves or bends. These at a distance appear like so many casks or hogsheads floating in a line with a considerable distance between them. Mr. Tuchsen, of Heroe, is the only person of the many correspondents I have, that informs me he has observed the difference between the body and the tail of this creature as to thickness. It appears that this creature does not, like the eel or land-snake, taper gradually to a point ; but the body, which looks to be as big as two hogsheads, grows remarkably small at once just where the tail begins. The whole animal is of a dark brown colour, but it is speckled and variegated with light streaks or spots, that shine like tortoise-shell. It seems the wind is so destructive to this creature, that, as has been observed before, it is never seen on the surface of the water, but in the greatest calm, and the least gust of wind drives him immediately to the bottom again. These creatures shoot through the water like an arrow out of a bow, seeking constantly the coldest places.

"If any one inquires how many folds may be counted in a Sea-Snake, the answer is, that the number is not always the same, but depends upon the various sizes of them ; *five and twenty* is the greatest number I find well attested."

A letter, dated Feb. 1751, is also quoted, from the Hon.

Lawrence de Ferry, a Captain in the navy, to Mr. John Reutz, Procurator, relating to a Sea-Snake, seen by him near MOLDE, on a very calm and hot day, in Aug. 1746. He had heard many stories about the monster—and ordered his men to row for it; he fired at it, on which it immediately sunk. Observing the water to be red, he supposed he had wounded it. The head, he relates, was like that of a horse—and of a grayish colour—the mouth was quite black and very large.—He also mentions the bright mane. The eyes were black, and there were seven or eight thick folds, about six feet distance from one another.

This letter, the substance of which we have given, was sworn to before the magistrates at Bergen.

The account had been published by Guthrie, in his geography, who gave credence to the story—but the editors of the Encyclopedia, after reciting the evidence, avow their disbelief of the existence of the Sea-Serpent—"Its bulk," say they, "is said to be so disproportionate (600 feet in length) to all the known animals of our globe, that it requires more than ordinary evidence to render it credible."

It is true the enormous length ascribed to the animal might reasonably have induced doubts of its accuracy; but they well knew the fallacy of determining by the eye the dimensions of objects at sea. A monstrous animal might yet have been seen, though his bulk should have been exaggerated. They did not credit the marvellous in regard to the length of the Snake, and disbelieved the whole relation. But what is very singular, the same editors express no disbelief of the story mentioned by Livy, of the enormous land Serpent, 120 *feet long*, which disputed the passage of the army of Regulus, over the river Bage da, in Africa. Now, this is four times the length of any land Serpent at present known, and its bulk might readily have awakened suspicions as to the veracity of the rest of the narration, if they had not chosen to believe a Roman historian, rather than a Danish bishop. But we see no absolute necessity for discrediting either the one or the other: there is probably an exaggerated statement in both.

The next most authentic account, as far as we are informed, of the appearance of a Sea-Serpent, has been published in the American Journal of Science, conducted by Professor Silliman, of Yale College.* In the year 1804, a letter was addressed by Alden Bradford, Esq to the Hon. John Q. Adams, Secretary to the American Academy of Arts and Sciences, transmitting several documents, tending to establish the fact, that a large Sea-Serpent had been seen in and near the Bay of Penobscot, at various times. Though the descriptions of the phenomenon were similar in character, yet the Academy thought them incredible, and did not make the statements public. The accounts were mislaid; but have since been recovered. The first is a letter from the Rev. Mr. A. Cummings, of Sullivan, dated August 17, 1803, to another clergyman, Mr. H. McLean, with some remarks by the latter. The second is another letter, dated August, 1804, from Mr. Cummings to Mr. Bradford.

The Sea-Serpent was seen by Mr. Cummings, his wife, daughter, and another lady, as they were on their passage to Belfast, between Cape Rosoi and Long-Island. It was in the month of July; the sea was calm; there was very little wind; and the first appearance of the Serpent was near Long-Island. Mr. C. supposed it to be a large shoal of fish, with a seal at one end of it; but he wondered *the seal should rise out of the water so much higher than usual ;* as he drew near, they discovered the whole appearance to be one animal in the form of a Serpent. *He had not the horizontal, but an ascending and descending serpentine mo-*

* Vol. 2d, No. 1. April 1820.

o5

tion. This account also refers to the descriptions given by other persons of similar animals seen, by people on Fox-Island, in Penobscot Bay, at different times, at Ash Point—at or near Boothbay—at Muscongus Bay—and off Meduncook.

There is also a letter from Capt. George Little, of the U. S. navy, now deceased, dated March, 1804, and is addressed to Mr. Bradford. In the year 1780, some time in May, he observes, as he was lying in Broad Bay (Penobscot,) in a public armed ship, he discovered at sunrise, a large Serpent, coming down the bay on the surface of the water. The cutter was manned and armed : he went himself in the boat ; and when within 100 feet of the Serpent, the marines were ordered to fire on him ; but before they could make ready, he plunged into the water. He was not less than 45 to 50 feet long : the largest diameter of his body he supposed to be 15 inches ; and his head, nearly the size of that of a man, he carried four or five feet out of water. He wore every appearance of a black Snake. He was afterwards pursued, but they never came nearer to him than a quarter of a mile. A Mr. Joseph Kent, of Marshfield, says Capt. Little, saw a like animal at the same place in the year 1751, which was longer and larger than the main boom of his sloop, of 85 tons. He observed him within ten or twelve yards of the vessel.

The declaration of Eleazer Crabtree is then given, who lived at Fox-Island, in the Bay of Penobscot, in the year 1777 or 1778. He had frequently heard of a sea-monster frequenting the waters near the shore—and doubting the tact, he went down one day upon receiving information from a neighbour, that he was then in the sea near his house. He saw a large animal in the form of a Snake, lying almost motionless in the water, about 500 feet from the bank where he stood. His head was about four feet above the surface ; he appeared 100 feet long; and he supposed him to be three feet in diameter. Many other inhabitants, upon whose veracity he could depend, had also declared to him that at other times they had seen such an animal.

The foregoing is the substance of the Documents published in the American Journal of Science.

A letter from Captain Crabtree, (whether he was the same person or not, we cannot say,) dated 1793, was published in the newspapers of the time, and afterwards reprinted. It related to a Serpent which he saw near Mount Desert, (Penobscot,) in the month of June, 1793. His head was elevated six or eight feet, somewhat resembling a horse's, and was larger than a barrel; the body was about the same circumference, of a dark brown colour, and appeared to be from 55 to 60 feet long. It showed no hostile disposition and remained in view of his vessel for nearly an hour. Its eyes were black and piercing, and its motion was rapid.

We are informed* on the authority of the Rev. Mr. A. Cummings, who was living in 1817, that the British, in their expedition to Bagaduse, had a view of the animal; though they exaggerated his length to 300 feet: that a Mr. Crocket saw *two* of them together about 25 years ago, and that the people on Mount Desert have also seen the animal. Various other accounts relative to the presence of this animal near the Penobscot Bay, in former years, are extant; but they are not sufficiently precise, to establish any further important facts, as to his size, motions and habits. All the narrations agree, however, with regard to his protuberances, his vertical sinuosities, his ser-

---

* In a letter from the Rev. William Jenks to the Hon. Judge Davis, dated Bath, Sept. 17, 1817—and published in the Report of a committee of the Linnæan Society of that year.

pent shaped head, and great magnitude. Upon this subject it may be proper to notice the large skeleton of an unknown animal which was found upon the shore of Mount
Desert, a number of years past.

The next appearance of this monster on our coast, we
believe, was about the 20th of June, in the year 1815, at
Warren's Cove, near Plymouth. He was seen by Mr. Elk-
nah Finney, a respectable mariner, who had always been
accustomed to foreign voyages and fishing—and had frequently seen whales and various species of large fish.
He deposed on oath, in August, 1817, that he first saw
something which appeared to the naked eye like drift sea-
weed. He then looked through a perspective glass, and
saw it was an aquatic animal, unknown to him. It was in
rapid motion northwardly, about a quarter of a mile from
the shore. At first it showed a length of about thirty feet
—but in turning, about half a mile off, it displayed at least
one hundred feet.. It afterwards came nearer, when it
stopped and lay entirely still on the surface for five minutes,
or more. The appearance, was like a string of buoys—
thirty or forty of which, of about the size of a barrel, were
exhibited. The head appeared to be six or eight feet long,
and where it was connected with the body was larger than
the body, but it tapered off to the size of a horse's head.*
The body was of a deep brown or black colour, but the
tail was not visible. The sea was calm, the wind light,
and the sky clear.

The next day he was seen again, by Mr. Finney and
others—but the observers could not determine whether his
motion was ascending and descending—or horizontal,
though it was very rapid. Certain house-carpenters, who
were at work near this cove, and other persons also, saw

---

* By the head is probably meant, the head and *neck*, as far as the
first protuberance.

the animal—who all mentioned with emphasis the long and distinct *wake* made in the water by the passage of his body through it.

The frequent visits made by this anomalous creature to the harbour of Gloucester, in the year 1817, have probably rendered him more notorious than any other circumstance connected with his history. The Linnæan Society have collected and published numerous facts upon the subject, taken from individuals who were separately examined, and who delivered their depositions under the strongest and most solemn pledge to secure veracity. Other individuals made personal inquiries respecting the animal; and the information they obtained we shall endeavour to generalize and convey in a narrow compass. It must be understood, that what we shall relate, will be a faithful summary of the testimony given by the deponents on oath, as well as of the reports of other respectable gentlemen.

The first information which the inhabitants of Gloucester received of the presence of the Serpent in their harbor, was obtained from the master of a coasting vessel, belonging to George's River, in Maine. He reported, early in August, that he had been frightened, by a huge Serpent laying along side of his vessel; his head appeared to be resting on the cable, and his tail extending beyond the stern. The vessel, according to her rate of tonnage, must have been about sixty feet long. The account was discredited, and the master sought refuge from ridicule on board of his sloop.—A subsequent relation given by Amos Story, a respectable man, living upon the extreme point of the harbour, excited the curiosity, though it was not calculated to remove the doubts of the inhabitants. On the 10th of August, he saw a strange animal, which he believed to be a Serpent, moving rapidly through the water, and elevating his head, which was shaped like that of a sea-turtle, about a foot above the surface. He continued in sight for an hour and a half. His colour was of a dark brown, and his circumference was about equal to that of a man's body. He (Mr. Story) saw fifty feet of his length, but did not discover any bunches upon his back.

During the month of August, he was seen by a variety of different individuals, eleven or twelve of whom observed him at distances, varying from forty to eight hundred feet. In most cases, when he was seen near, he did not show his body more than from twelve to twenty feet out of water— in which instances the protuberances were not evident; but those persons who saw 60, 70 or 90 feet of his length, all mention his joints, rings, bunches or swells; and one gentleman in particular, who saw him lying still, observed these bunches very distinctly, about one foot in height, upon his back. Capt. Tappan, and two of his crew, on board the Laura, of Newburyport, saw his head within 30 or 50 feet, and describe it with minuteness. It was formed like that of a Serpent's; his *tongue* was thrust out, and appeared about two feet in length: this he raised several times over his head, and then let it fall again: it was of a light brown colour, and the end of it resembled a harpoon. The eye was like that of an ox, and there appeared to be a small bunch over it on each side of his head. The animal did not appear to be disturbed by the vessel, and his motion was much swifter than that of any whale. When he contracts his joints in a propelling motion, his folds appear rugged and bristling with life; but when he ceases to move, his scales enclose, and he seems to be comparatively smooth. He swims underneath the water with greater rapidity, than when his body is exposed to view upon the surface. The head has been described as a terrific object; its dark brown colour is mixed with some streaks of a lighter hue, and it is likened in its hard ⁓d scaly appearance to a " weather-beaten rock." One person, Mr. Gaffeny, at Gloucester, a good marksman, fired at him with his gun at thirty feet distance, with a ball eighteen to the pound, which he supposed must have taken effect, though the ball might, and probably did turn aside, in consequence of glancing upon his scales.

Besides those persons who gave their testimony upon oath, there were many who were eye-witnesses of his appearance.

Capt. Obear, arrived at Beverly, after having put into Cape Ann. He reported his astonishment at the sight of a monstrous animal, in the form of a Serpent, lying upon the water near that harbor. Kettle-cove, at Manchester, was a favourite place of resort for the Serpent, and he was often seen on the 21st and 22d of August. Capt. Burchmore, who arrived at Salem from New-York, observed the animal near Half-way Rock, on the 6th of September; and next day afternoon he frequently approached so near to several fishing-boats, off the harbour of Gloucester, that the people were frightened: he seemed to be employed in seeking food, and was once, according to their report, within ten feet of them. He was also reported to have been seen near Wells, (Kennebunk) on the 12th of September.*

---

* A great deal of excitement was occasioned in Boston, by the return of an expedition (which went out with the view of capturing the Sea-Serpent), with a fish called a *Horse Mackerel.* No small ridicule was, in consequence, cast upon the believers in the real existence of the Serpent; and many were converted, by the event, from belief to infidelity. But it is easy to perceive that this matter has no real bearing upon the question, further than this :—it shows that the captors believed the Sea-Serpent to be only a horse mackerel. The ridicule would seem to be rather on the side of those who maintain that so insignificant a fish is capable of exhibiting the various extraordinary appearances attested to by hundreds of witnesses, and ascribed to the Sea-Serpent.

The Sea-Serpent made his appearance again in 1818. The following is the deposition of a gentleman well known for his correctness and veracity.   It was first published in the *Hallowell Advocate* (Me).

"HALLOWELL, JUNE 27, 1818.

"I, Shubael West, of Hallowell, in the County of Kennebec, master of the Packet Delia, plying between Kennebec river and Boston, testify and say that I left Boston on the morning of Sunday the 21st inst.; and at about 6 o'clock, P. M., Cape Ann bearing W. S. W. about 2 leagues, steering a course N. N. E., saw directly ahead, distant three quarters of a mile, an object, which I have no doubt was the Sea-Serpent so often mentioned by others, engaged with a whale, that was endeavouring to elude the attack.   The Serpent threw up his tail from 25 to 30 feet in a perpendicular direction; striking the whale with tremendous blows, rapidly repeated, which were distinctly heard and very loud, for two or three minutes.   They then both disappeared for several minutes, moving in a W. S. W. direction, when they reappeared in shore of us, and about under the sun, the reflection of which was so strong as to prevent our seeing so distinctly as before,—when the tremendous blows were repeated and as clearly heard as before.   They again went down for a short time, and again came up to the surface under our larboard quarter, the whale appearing first, and the Serpent in pursuit. Here our view was very fine.   The Serpent shot up his tail through the water to the height before mentioned, which he held out of water some time, waving it in the air; and at the same time, while his tail remained in this position, raised his head rather leisurely 15 or 20 feet, as if taking a view of the surface of the sea.   After remaining in this situation a short time, he again sunk into the water, and disappeared.

"The Serpent's body was larger, in my opinion, than the

mast of any ship I ever saw. His tail appeared very rag-
ged and rough, and was shaped something like an eel's;
and his head like that of the land Serpent's. Being well
acquainted with whaling, I think the whale was endeavouring to escape, as he spouted but once at a time on coming
to the surface. The whale's back was distinctly seen, as
well as his spouting, by 15 or 18 persons, as well as myself, with the exception of one woman.

SHUBAEL WEST."

This statement is confirmed by a very long and minute
affidavit of Mr. Samuel Schmid, from which we shall se-
lect such passages as are most interesting. Professor
Cooper, of the University of Pennsylvania, says of this
gentleman, "I have known and highly respected Mr.
Schmid for many years, and I consider his character for
probity and veracity as unimpeachable. Nor is it easy to
find a man of more attentive and accurate observations,
&c. * * * An account of a scene of which he was an eye-
witness, given me by a gentleman of this description,
seemed to me entitled to more than the usual credit given
to strange occurrences, related by persons whose characters are little known, &c. &c." These documents first
appeared in the *Hallowell Gazette*.

"I, Samuel Schmid, of Philadelphia, testify and say as
follows:

"On Sunday, the 21st inst., between the hours of 8 and
6, P. M. about 2 or 3 leagues to the E. N. E. of Cape Ann,
being on board the Delia, Captain Shubael West, on my
passage from Boston to Hallowell in Maine, I saw, ahead
of the vessel, a tall white object, standing upright out of
the water, which I thought might be a pillar set up for
some particular purpose. I viewed it alone therefore for
7 or 8 minutes, till finally I saw its upper end waving
about for a short time; when, after giving a hard stroke
to the water, it disappeared. In about 3 or 4 minutes this

object rose again: when I called upon various persons below in the vessel, to come up and view it, which they did immediately. At this time a whale appeared spouting near it; and the object which stood up, was after a time recognized as being the tail of some animal, and at length was concluded to be that of the noted Sea-Serpent. The tail was now seen to strike the water again several times. Both these great objects now went below the surface of the sea. In a few minutes both showed themselves again, abreast of us: but the sun lying behind both, and there being no glass at hand, our view of them grew imperfect. But soon afterwards, we perceived a form, like that of a head, rising up: the parts below which seemed connected with the tail which I had originally seen; though the intermediate body was to us invisible, the neck was curved below the head; the head was at first held horizontally; but afterwards assumed an oblique position, as if looking down into the water below. Some additional strokes were now made with the tail, the whale on its side now was in motion again, and immediately dived below, as did the Serpent. Some considerable time afterwards both animals appeared again, but at too great a distance for a perfect view of them to be had: but the Serpent made several fresh strokes. * * *. The motion of these animals were rapid, and those of the Sea-Serpent vigorous, the noise and the agitation of the water from his strokes being great. His tail, which I have said seemed white, appeared to be flattened crosswise; its edges also had an indented appearance, while its end was blunt; the head and neck appeared of a dark colour, and the body below the neck seemed rapidly to enlarge. I had an opportunity soon after to observe the mast of a vessel, which appeared abundantly smaller. The length of tail exhibited, I thought was about 25 or 30 feet, and the head stood about half this height above the water. The separation between the two seemed so considerable, that it was thought that our vessel might have sailed between them across the body. * * * Supposing the body to have been long like that of a snake, the total length of the animal may perhaps be estimated at 100 feet. Having lived in parts of the U. S. where the former accounts given of this Serpent have received little credit, I was not at first prepared to expect what I afterwards had the good fortune to see. I state mere facts, that those who know me may be assured of them, and also that naturalists may, from the statement given, begin to have some notion of the habits of this animal. * * * The persons on board who became companions with me in this sight, were two sea-captains besides Capt. West, with various passengers, and the crew." [Sworn to before John Merrick, Justice of the Peace.]

The Sea-Serpent was also seen off Cape Ann this season (1818). On one occasion, when he was encountered by Captain Webber and others, two harpoons were thrown at and struck him, but without making any incision. Several balls were also fired at him, with no effect but making him dive under water. On another of his appearances in the harbour, a boat went off with muskets, and they fired at him 7 or 8 times, but knew not whether they hit him : apparently they did not hurt him, however; for though on the firing he went under water, he soon rose again, and played about upon the surface as before.

We shall here introduce the statement of Mr. Hodgkins, an intelligent man of about 50 years of age, who had followed the sea most of his life.

"GLOUCESTER, AUG. 18, 1818.

"Returning from Newburyport into Squam harbour, on Wednesday last, in a chebacco boat, where we had been for the purpose of obtaining fishing-bait; and having failed to get a supply, we were in hopes of taking some on our passage. When off Chebacco bar, it being perfectly calm, we discovered something, at the distance of a mile or more, which we were in hopes was a shoal of bait, making a great agitation on the smooth surface of the water. It seemed to approach us rapidly; when it came nearer, we were convinced we had been in an error, and what we took for a shoal of black fish was nothing less than the bunches on the back of the celebrated Sea-Serpent. He made directly for the boat until he came within 50 yards: he then sunk under water, and we were much alarmed lest he should rise under us, as we had no power of getting from him, we lying becalmed. When he came up he was thirty feet from us, and we had a perfect view of him. His head was elevated from 3 to 5 feet; the distance about 6 feet from his neck to his first bunch. We counted 20 bunches, and supposed them on an average about 5 feet apart. When we first saw him there appeared a rippling in the water, which made a noise not unlike water running rapidly over loose pebbles: on his near approach we found it was the ripple made by the bunches on his back. It was 12 o'clock, noon, when we saw him; the weather clear and the sea smooth. His head was of a dark brown colour, formed like a seal's, and had a glossy appearance. His body was of the size of a 60 or 80 gallon cask."

During the summer of 1819, and especially during the month of August, the Serpent was seen more frequently than ever before, and by large numbers of people of the most unquestionable credit. The following is one among a multitude of statements.*

            "BROOKLINE, AUG. 19, 1819.

"I got into my chaise about 7 o'clock in the morning to come to Boston, and on reaching the long beach, observed a number of people collected there, and several boats pushing off, and in the offing. * * As my curiosity was

---

        * Columbian Centinel, Aug. 28, 1819.

directed towards the boats, to ascertain the course they were taking, my attention was suddenly arrested by an object emerging from the water at the distance of about 100 or 150 yards, which gave to my mind, at the first glance, the idea of a horse's head. As my eye ranged along, I perceived, at a short distance, eight or ten regular bunches or protuberances, and at a short interval 3 or 4 more. I was now satisfied that the Sea-Serpent was before me; and after the first moment of excitement produced by the sight of so strange a monster, tasked myself to investigate his appearance as accurately as I could.

" My first object was the head, which I satisfied myself was serpent shaped. It was elevated about 2 feet from the water, and he depressed it gradually to within 6 or 8 inches as he moved along. I could always see under his chin, which appeared to hollow underneath, or to curve downward. His motion was at that time very slow along the beach, inclining towards the shore: he at first moved his head from side to side, as if to look about him. I did not see his eyes, though I have no doubt I could have seen them if I had thought to attend to this. His bunches appeared to me not altogether uniform in size; and as he moved along, some appeared to be depressed, and others brought above the surface, though I could not perceive any motion in them. My next object was to ascertain his length: for this purpose I directed my eye to several whale-boats at about the same distance, one of which was beyond him, and by comparing the relative length, I calculated that the distance from the animal's head to the last protuberance I had noticed would be equal to about five of those boats. I felt persuaded by this examination that he could not be less than eighty feet long; and as he approached the shore, and came between me and a point of land which projects from the eastern part of the beach, I had another means of satisfying myself on this point.

" After I had viewed him thus attentively for about 4 or 5 minutes, he sunk gradually into the water and disappeared: he afterwards again made his appearance for a moment at a short distance.

" My first reflection after the animal was gone, was, that the idea I had received from the description you gave of the animal you saw at Gloucester in 1817, was perfectly realized in this instance; and that I had discovered nothing which you had not before described. The most authentic testimony given of his first appearance there seemed to me remarkably correct: and I felt as if the appearance of this monster had been already familiar to me.

" After remaining some 2 or 3 hours on the beach without again seeing him, I returned towards Nahant; and in crossing the small beach had another good view of him, but at a greater distance. At this time he moved more rapidly, causing a white foam under the chin, and a long wake, and his protuberances had a more uniform appearance. At this time he must have been seen by 2 or 300 persons on the beach, and on heights each side; some of whom were very favourably situated to observe him. * *"

This description is from the pen of Samuel Cabot, and is addressed to Col. T. H. Perkins, of Boston. The authenticity of this account we shall not attempt to strengthen: but other interesting particulars of the same appearance might be selected from a large mass of unquestionable testimony. The following is extracted from a letter of James Prince, Esq. U. S. marshal of the District of Boston, to the Hon. Judge Davis, of that city, dated Aug. 16, 1819. The scene described, took place at Nahant.

"His head appeared about 3 feet out of water. I counted 13 bunches on his back—my family thought there were 15. He crossed three times at a moderate rate across the bay, but so fleet as to occasion a foam in the water. My family and self, who were in a carriage, judged that he was from 50 and not more than 60 feet in length. * * As he swam up the bay, we and the other spectators moved on and kept nearly abreast of him. * * I had seven distinct views of him from the long beach, and at some of them the animal was not more than 100 yards distant. * * On passing the second beach, we were again gratified beyond even what we saw in the other bay, which I concluded he had left in consequence of the number of boats in the offing in pursuit of him. * * We had here more than a dozen different views of him, each similar to the other: one however so near that the coachman exclaimed—"See his glistening eye." * * I have been accustomed to see whales, sharks, grampuses, porpoises, &c; but he partook of none of the appearances of either of these. * * The water was extremely smooth, and the weather clear. The time occupied was from a quarter past eight to a quarter past eleven."

Those who are disposed to go further into the minutiæ of this subject, will find an abundance of testimony, of the most satisfactory nature, in the papers of 1819. The Columbian Centinel of Aug. 18, 1819, gives the statement of the Hon. Jas. T. Austin. The Boston Evening Gazette of June 12, 1819, gives two very minute and distinct depositions; and the same paper of the 14th Aug. furnishes various depositions, and statements authenticated by Alden Bradford, Esq.

It is unnecessary to add anything further on this subject, except the following notice of the most recent appearance of the Sea-Serpent in our waters.

From the Kennebunk Gazette of July, 1830.

"The coast in our immediate vicinity has at last received a visit from the far famed *Sea-Serpent.* He was seen by three men, who were fishing a few miles distant from the shore, on Thursday afternoon last. Two of the men were so much alarmed at his nearness to the boat, that they went below. The third, however, Mr. Gooch, a man whose statements can be relied on, remained on deck and returned the glances of his Serpentship for a considerable length of time. He gives the following account of the interview: The fish was first seen a short distance from them, and shortly after he turned about and came within *six feet* of the boat, when he raised his head about four feet from the water, and looked directly into the boat, and so remained for several minutes. Mr. Gooch noticed him attentively, and thinks he was sixty feet in length, and about six in circumference," &c.

———

We are perfectly aware of the fact that many persons totally disbelieve the existence of the Sea-Serpent, or an animal resembling it, as described in the preceding pages. We do not hesitate however to declare our faith in these statements. There is nothing in them either incredible or difficult to believe. The testimony is unimpeachable in its character, and abundant in amount. One half the undisputed materials of history rests upon a far slighter basis. It must be remarked too, that this mass of positive evidence is confronted by none of an opposite nature. We see no alternative, between receiving and believing the story, or assuming the position that some hundreds of persons, including many of our most respectable citizens, have combined, without any visible motive, to palm off a gross deception on the public; or have themselves been cheated by illusions more extraordinary than any recorded in history. It is hardly necessary to add that the credulity which embraces either of these latter conclusions, is much greater than that which yields to the evidence and believes in the existence of the Sea-Serpent.

CPSIA information can be obtained
at www.ICGtesting.com
Printed in the USA
BVHW04*1020260718
522720BV00008B/88/P

9 780365 039730